National Qualifications 2015

X707/75/02

Biology
Section 1—Questions

WEDNESDAY, 13 MAY

9:00 AM – 11:00 AM

Instructions for the completion of Section 1 are given on Page two of your question and answer booklet X707/75/01.

Record your answers on the answer grid on Page three of your question and answer booklet

Before leaving the examination room you must give your question and answer booklet to the Invigilator; if you do not, you may lose all the marks for this paper.

X7077502

SECTION 1

1. In the diagrams below, the circles represent molecules on either side of a cell membrane. In which of these diagrams would the molecules move into a cell by diffusion?

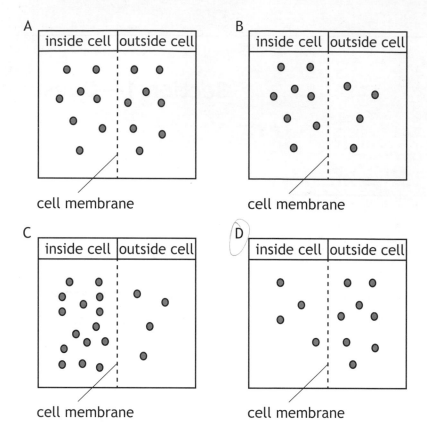

2. Which of the following does **not** involve mitosis?

   A   Synthesis of proteins

   B   Growth of tissue

   C   Maintenance of the diploid chromosome complement

   D   Repair of tissue

3. The graph below shows changes in the enzyme and substrate concentrations in a seed over a period of time.

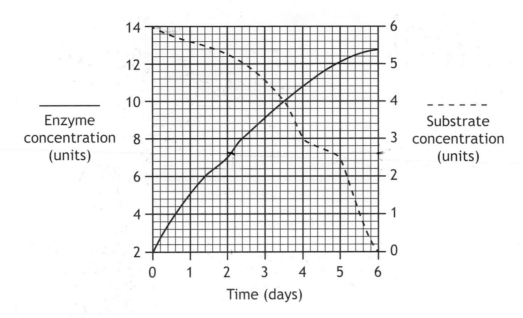

Enzyme concentration (units) _____

Substrate concentration (units) - - - - - -

Time (days)

How many days does it take for the substrate concentration to decrease by 50%?

A   2

B   3

C   4

D   5

4. Some stages of genetic engineering are shown below.

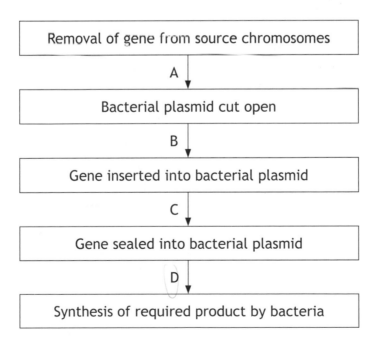

Which letter indicates the stage where the plasmid is inserted into a bacterial cell?

5. The effect of light intensity on the rate of photosynthesis was measured for two species of plants, L and M.

    The results are shown in the graph below.

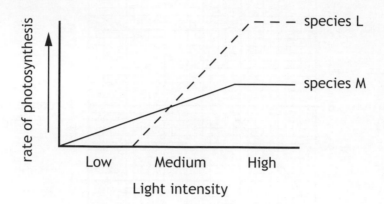

    The rate of photosynthesis of species M is

    A   slower than L in low light intensities

    B   slower than L in high light intensities

    C   faster than L in medium light intensities

    D   faster than L in high light intensities.

6. The diagrams below show four different types of cell.

    Which cell was produced by a meristem?

    A    B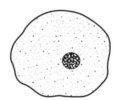

    C 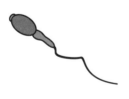   D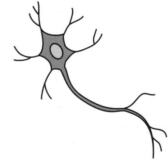

7. The diploid number of chromosomes in a cell from a kangaroo is 12.

Which line in the table below identifies the number of chromosomes for the cell type shown?

	Kangaroo Cell Type	Number of chromosomes
A	sperm	12
B	skin	6
C	nerve	6
D	zygote	12

8. The diagrams below show the same sections of matching chromosomes found in four flies, A, B, C and D.

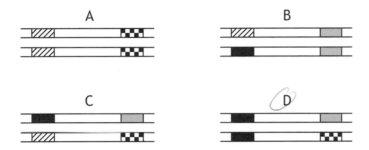

The alleles shown on the chromosomes can be identified using the following key.

▨ allele for striped body

■ allele for unstriped body

▢ allele for normal antennae

▩ allele for abnormal antennae

Which fly is homozygous for body pattern and heterozygous for antennae type?

[Turn over

9. The diagram below shows an alveolus and an associated blood capillary.

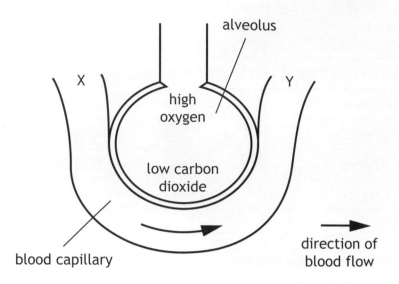

As blood flows from X to Y gases are exchanged with the alveolus.

Which line in the table below identifies the concentrations of gases at X and Y?

	Concentration at X	Concentration at Y
A	high oxygen	high carbon dioxide
B	low oxygen	high carbon dioxide
C	low oxygen	low carbon dioxide
D	high oxygen	low carbon dioxide

10. The following sequence shows part of the blood flow through the body.

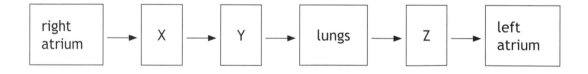

Which line in the table below identifies X, Y and Z?

	X	Y	Z
A	right ventricle	pulmonary vein	pulmonary artery
B	right ventricle	pulmonary artery	pulmonary vein
C	pulmonary vein	pulmonary artery	right ventricle
D	pulmonary artery	right ventricle	pulmonary vein

11. The graph below shows the relationship between the concentration of carbon dioxide and oxyhaemoglobin in the blood.

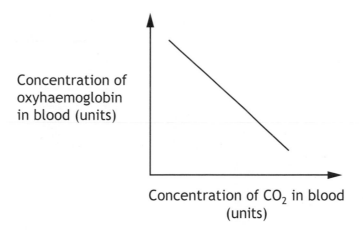

Which of the following statements describes this relationship?

A   As the carbon dioxide concentration increases the concentration of oxyhaemoglobin decreases.

B   As the carbon dioxide concentration decreases the concentration of oxyhaemoglobin decreases.

C   As the carbon dioxide concentration increases the concentration of oxyhaemoglobin increases.

D   Increasing carbon dioxide concentration has no effect upon the concentration of oxyhaemoglobin.

[Turn over

12. The chart below shows the percentage of men and women with obesity at different ages, in a population.

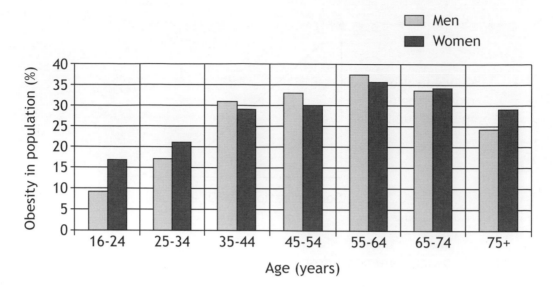

Which of the following statements is true?

A   For each age group there is a higher percentage of obese men than obese women.

B   For each age group there is a higher percentage of obese women than obese men.

C   Obesity in men and women increases with age up to 64 years.

D   Obesity in men and women decreases with age up to 64 years.

13. Which of the following statements best describes a biome?

A   All the organisms in an area and their habitat.

B   The role that an organism plays within a community.

C   A living factor which affects biodiversity in an ecosystem.

D   A region of our planet as distinguished by its climate, fauna and flora.

14. The size of a population of snails can be estimated using the following formula.

$$\text{Population} = \frac{\text{Number collected on 1st day} \times \text{Number collected on 2nd day}}{\text{Number of marked individuals found on 2nd day}}$$

A student investigated the population of snails in a garden. He collected 40 snails, marked their shells and released them. Next day, 35 snails were collected and 14 of these were found to be marked.

The snail population was estimated to be

A   16

B   100

C   560

D   1400.

15. Which of the following describes interspecific competition?

    A   Individuals of different species requiring different resources.

    B   Individuals of different species requiring similar resources.

    C   Individuals of the same species requiring different resources.

    D   Individuals of the same species requiring similar resources.

16. The diagram below represents four populations of animals P, Q, R and S and areas of interbreeding. Interbreeding takes place in the shaded areas.

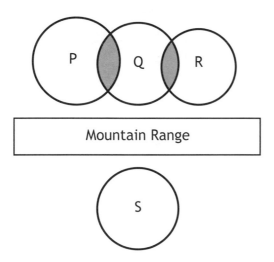

How many species may evolve over time?

    A   1

    B   2

    C   3

    D   4

17. Antibiotic resistance in bacteria is an example of evolution. Which of the following shows the sequence of events leading to this?

    A   Natural selection ⟶ mutation ⟶ use of antibiotic

    B   Mutation ⟶ natural selection ⟶ use of antibiotic

    C   Mutation ⟶ use of antibiotic ⟶ natural selection

    D   Natural selection ⟶ use of antibiotic ⟶ mutation

[Turn over

18. The graph below shows information about the growth of the human population.

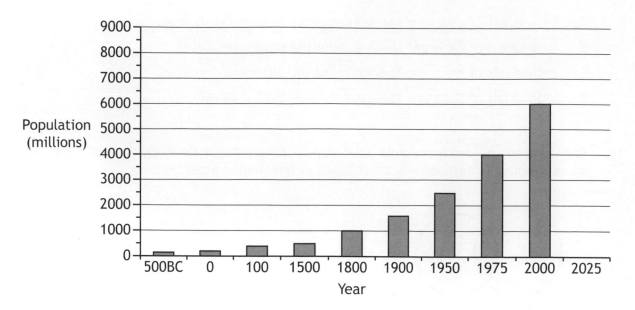

If the population continues to increase at the same rate as between 1975 and 2000, predict the population size in 2025.

A   8000

B   7500

C   8500

D   9000

19. DDT can be sprayed onto crops to kill insects. It can be washed off the crops by rainwater and flow into rivers where it accumulates in food chains.

A typical freshwater food chain and the concentration of DDT in each organism is shown below.

Food chain: algae ⟶ stickleback ⟶ trout ⟶ osprey

DDT concentration: 0·001   2·0   5·0   20·0

The percentage increase in DDT concentration between the trout and osprey is

A   15

B   100

C   300

D   400.

20. Which of the following statements describes the sequence of events when fertiliser leaches into a loch?

    A   Algal bloom develops ⟶ algae die ⟶ oxygen concentration increases

    B   Algal bloom develops ⟶ algae die ⟶ oxygen concentration decreases

    C   Oxygen concentration increases ⟶ algal bloom develops ⟶ algae die

    D   Algae die ⟶ oxygen concentration decreases ⟶ algal bloom develops

[END OF SECTION 1. NOW ATTEMPT THE QUESTIONS IN SECTION 2 OF YOUR QUESTION AND ANSWER BOOKLET]

[BLANK PAGE]

DO NOT WRITE ON THIS PAGE

IN ASSOCIATION WITH

SQA

Hodder Gibson
Model Practice
Papers
WITH ANSWERS

PLUS: Official SQA Specimen Paper & 2014 Past Paper With Answers

National 5
Biology

2013 Specimen Question Paper, Model Papers & 2014 Exam

HODDER
GIBSON
AN HACHETTE UK COMPANY

This book contains the official 2013 SQA Specimen Question Paper and 2014 Exam for National 5 Biology, with associated SQA approved answers modified from the official marking instructions that accompany the paper.

In addition the book contains model practice papers, together with answers, plus study skills advice. These papers, some of which may include a limited number of previously published SQA questions, have been specially commissioned by Hodder Gibson, and have been written by experienced senior teachers and examiners in line with the new National 5 syllabus and assessment outlines, Spring 2013. This is not SQA material but has been devised to provide further practice for National 5 examinations in 2014 and beyond.

Hodder Gibson is grateful to the copyright holders, as credited on the final page of the Answer Section, for permission to use their material. Every effort has been made to trace the copyright holders and to obtain their permission for the use of copyright material. Hodder Gibson will be happy to receive information allowing us to rectify any error or omission in future editions.

Hachette UK's policy is to use papers that are natural, renewable and recyclable products and made from wood grown in sustainable forests. The logging and manufacturing processes are expected to conform to the environmental regulations of the country of origin.

Orders: please contact Bookpoint Ltd, 130 Park Drive, Abingdon, Oxon OX14 4SE. Telephone: (44) 01235 827720. Fax: (44) 01235 400454. Lines are open 9.00–5.00, Monday to Saturday, with a 24-hour message answering service. Visit our website at www.hoddereducation.co.uk. Hodder Gibson can be contacted direct on: Tel: 0141 848 1609; Fax: 0141 889 6315; email: hoddergibson@hodder.co.uk

This collection first published in 2014 by
Hodder Gibson, an imprint of Hodder Education,
An Hachette UK Company
2a Christie Street
Paisley PA1 1NB

BrightRED   Hodder Gibson is grateful to Bright Red Publishing Ltd for collaborative work in preparation of this book and all SQA Past Paper, National 5 and Higher for CfE Model Paper titles 2014.

Typeset by PDQ Digital Media Solutions Ltd, Bungay, Suffolk NR35 1BY

Printed in the UK

A catalogue record for this title is available from the British Library

ISBN: 978-1-4718-3695-4

3 2 1

2015 2014

Introduction

Study Skills – what you need to know to pass exams!

Pause for thought

Many students might skip quickly through a page like this. After all, we all know how to revise. Do you really though?

Think about this:

"IF YOU ALWAYS DO WHAT YOU ALWAYS DO, YOU WILL ALWAYS GET WHAT YOU HAVE ALWAYS GOT."

Do you like the grades you get? Do you want to do better? If you get full marks in your assessment, then that's great! Change nothing! This section is just to help you get that little bit better than you already are.

There are two main parts to the advice on offer here. The first part highlights fairly obvious things but which are also very important. The second part makes suggestions about revision that you might not have thought about but which WILL help you.

Part 1

DOH! It's so obvious but …

Start revising in good time

Don't leave it until the last minute – this will make you panic.

Make a revision timetable that sets out work time AND play time.

Sleep and eat!

Obvious really, and very helpful. Avoid arguments or stressful things too – even games that wind you up. You need to be fit, awake and focused!

Know your place!

Make sure you know exactly **WHEN and WHERE** your exams are.

Know your enemy!

Make sure you know what to expect in the exam.

How is the paper structured?

How much time is there for each question?

What types of question are involved?

Which topics seem to come up time and time again?

Which topics are your strongest and which are your weakest?

Are all topics compulsory or are there choices?

Learn by DOING!

There is no substitute for past papers and practice papers – they are simply essential! Tackling this collection of papers and answers is exactly the right thing to be doing as your exams approach.

Part 2

People learn in different ways. Some like low light, some bright. Some like early morning, some like evening / night. Some prefer warm, some prefer cold. But everyone uses their BRAIN and the brain works when it is active. Passive learning – sitting gazing at notes – is the most INEFFICIENT way to learn anything. Below you will find tips and ideas for making your revision more effective and maybe even more enjoyable. What follows gets your brain active, and active learning works!

Activity 1 – Stop and review

Step 1

When you have done no more than 5 minutes of revision reading STOP!

Step 2

Write a heading in your own words which sums up the topic you have been revising.

Step 3

Write a summary of what you have revised in no more than two sentences. Don't fool yourself by saying, "I know it, but I cannot put it into words". That just means you don't know it well enough. If you cannot write your summary, revise that section again, knowing that you must write a summary at the end of it. Many of you will have notebooks full of blue/black ink writing. Many of the pages will not be especially attractive or memorable so try to liven them up a bit with colour as you are reviewing and rewriting. **This is a great memory aid, and memory is the most important thing.**

Activity 2 — Use technology!

Why should everything be written down? Have you thought about "mental" maps, diagrams, cartoons and colour to help you learn? And rather than write down notes, why not record your revision material?

What about having a text message revision session with friends? Keep in touch with them to find out how and what they are revising and share ideas and questions.

Why not make a video diary where you tell the camera what you are doing, what you think you have learned and what you still have to do? No one has to see or hear it, but the process of having to organise your thoughts in a formal way to explain something is a very important learning practice.

Be sure to make use of electronic files. You could begin to summarise your class notes. Your typing might be slow, but it will get faster and the typed notes will be easier to read than the scribbles in your class notes. Try to add different fonts and colours to make your work stand out. You can easily Google relevant pictures, cartoons and diagrams which you can copy and paste to make your work more attractive and **MEMORABLE**.

Activity 3 – This is it. Do this and you will know lots!

Step 1

In this task you must be very honest with yourself! Find the SQA syllabus for your subject (www.sqa.org.uk). Look at how it is broken down into main topics called MANDATORY knowledge. That means stuff you MUST know.

Step 2

BEFORE you do ANY revision on this topic, write a list of everything that you already know about the subject. It might be quite a long list but you only need to write it once. It shows you all the information that is already in your long-term memory so you know what parts you do not need to revise!

Step 3

Pick a chapter or section from your book or revision notes. Choose a fairly large section or a whole chapter to get the most out of this activity.

With a buddy, use Skype, Facetime, Twitter or any other communication you have, to play the game "If this is the answer, what is the question?". For example, if you are revising Geography and the answer you provide is "meander", your buddy would have to make up a question like "What is the word that describes a feature of a river where it flows slowly and bends often from side to side?".

Make up 10 "answers" based on the content of the chapter or section you are using. Give this to your buddy to solve while you solve theirs.

Step 4

Construct a wordsearch of at least 10 X 10 squares. You can make it as big as you like but keep it realistic. Work together with a group of friends. Many apps allow you to make wordsearch puzzles online. The words and phrases can go in any direction and phrases can be split. Your puzzle must only contain facts linked to the topic you are revising. Your task is to find 10 bits of information to hide in your puzzle, but you must not repeat information that you used in Step 3. DO NOT show where the words are. Fill up empty squares with random letters. Remember to keep a note of where your answers are hidden but do not show your friends. When you have a complete puzzle, exchange it with a friend to solve each other's puzzle.

Step 5

Now make up 10 questions (not "answers" this time) based on the same chapter used in the previous two tasks. Again, you must find NEW information that you have not yet used. Now it's getting hard to find that new information! Again, give your questions to a friend to answer.

Step 6

As you have been doing the puzzles, your brain has been actively searching for new information. Now write a NEW LIST that contains only the new information you have discovered when doing the puzzles. Your new list is the one to look at repeatedly for short bursts over the next few days. Try to remember more and more of it without looking at it. After a few days, you should be able to add words from your second list to your first list as you increase the information in your long-term memory.

FINALLY! Be inspired...

Make a list of different revision ideas and beside each one write **THINGS I HAVE** tried, **THINGS I WILL** try and **THINGS I MIGHT** try. Don't be scared of trying something new.

And remember – "FAIL TO PREPARE AND PREPARE TO FAIL!"

National 5 Biology

The Course

The National 5 Biology Course consists of three National Units. These are *Cell Biology*, *Multicellular Organisms* and *Life on Earth*. In each of the Units you will be assessed on your ability to demonstrate and apply knowledge of Biology, and to demonstrate and apply skills of scientific inquiry. Candidates must also complete an Assignment in which they research a topic in biology and write it up as a report. They also take a Course Examination.

How the Course is graded

To achieve a Course award for National 5 Biology you must pass all three National **Unit Assessments**, which will be assessed by your school or college on a pass or fail basis. The grade you get depends on the following two Course assessments, which are set and graded by SQA.

1 **An Assignment** that requires you to write a 500 – 800 word report. The Assignment is 20% of your grade and is marked out of 20 marks, most of which are allocated for skills of scientific inquiry.

2 A written **Course Examination**, which is worth the remaining 80% of the grade. The Examination is marked out of 80 marks, most of which are for the demonstration and application of knowledge, although there are also marks available for skills of scientific inquiry. This book should help you practise the Examination part!

To pass National 5 Biology with a C grade you will need about 50% of the 100 marks available for the Assignment and the Course Examination combined. For a B, you will need 60%, and for an A, 70%.

The Course Examination

The Course Examination is a single question paper split into two sections. The first section is an objective test with 20 multiple choice items for 20 marks. The second section is a mixture of restricted and extended response questions worth between 1 and 3 marks each for a total of 60 marks. Some questions will contain options and there will usually be a question that asks you to suggest changes to experimental methods. Altogether there are 80 marks, and you will have 2 hours to complete the paper. Most of the marks are for knowledge and its application, with the remainder of questions designed to test skills of scientific inquiry.

The majority of the marks will be straightforward – these are the marks that will help you get a grade C. Some questions will be more demanding – these are the questions you need to get right to get a grade A.

General hints and tips

You should have a copy of the Course Assessment Specification for National 5 Biology – if you haven't got one, download it from the SQA website. This document tells you what you may be tested on in your examination. It is worth spending some time studying this document.

This book contains four practice National 5 Biology examination papers. One is the SQA specimen paper and there are three model papers. Notice how similar they all are in the way in which they are laid out and the types of question they ask – your own Course Examination will be very similar as well so working through the papers in this book will be good preparation.

If you are trying a whole examination paper from this book, give yourself a maximum of two hours to complete it. The questions in each paper are laid out in Unit order. Make sure that you spend time using the answer section to mark your own work – it is especially useful if you can get someone to help you with this. You could even grade your work on an A–D basis.

The following hints and tips are related to examination techniques as well as avoiding common mistakes.

Remember that if you hit problems with a question, you should ask your teacher for help.

Section 1

20 multiple-choice items   20 marks

- Answer on a grid.
- Do not spend more than **30 minutes** on this section.
- Some individual questions might take longer to answer than others – this is quite normal and make sure you use scrap paper if a calculation or any working is needed.
- Some questions can be answered instantly – again, this is normal.
- **Do not leave blanks** – complete the grid for each question as you work through.
- Try to answer each question in your head **without** looking at the options. If your answer is there – you are home and dry!

- If you are not certain, choose the answer that seemed most attractive on **first** reading the answer options.

- If you are guessing, try to eliminate options before making your guess. If you can eliminate 3 – you are left with the correct answer even if you do not recognise it!

Section 2

Restricted and extended response   60 marks

- Spend about **90 minutes** on this section.

- Answer on the question paper. Try to write neatly and keep your answers on the support lines if possible – the lines are designed to take the full answer!

- A clue to answer length is the mark allocation – most questions are restricted to 1 mark and the answer can be quite short. If there are 2 or 3 marks available, your answer will need to be extended and may well have two, three or even four parts.

- The questions are usually laid out in Unit sequence but remember some questions are **designed** to cover more than one Unit.

- The grade C-type questions usually start with "**State**", "**Identify**", "**Give**" or "**Name**" and often need only a word or two in response. They will usually be worth one mark each.

- Questions that begin with "**Explain**" and "**Describe**" are usually grade A types and are likely to have more than one part to the full answer. You will usually have to write a sentence or two and there may be two or even three marks available.

- Make sure you read questions through twice before trying to answer – there is often very important information within the question.

- Using abbreviations like DNA and ATP is fine, and the bases of DNA can be given as A, T, G and C.

- Don't worry that a few questions are in unfamiliar contexts – that's the idea! Just keep calm and read the questions carefully.

- If a question contains a choice, be sure to spend a minute or two making the best choice for you.

- In experimental questions, you must be aware of what variables are, why controls are needed and how reliability might be improved. It is worth spending time on these ideas – they are essential and will come up year after year.

- Some candidates like to use a highlighter pen to help them focus on the essential points of longer questions – this is a great technique.

- Remember that a **conclusion** can be seen from data, whereas an **explanation** will usually require you to supply some background knowledge as well.

- Remember to "**use values from the graph**" when describing graphical information in words if you are asked to do so.

- Plot graphs carefully and join the plot points using a ruler. Include zeros on your scale where appropriate and use the data table headings for the axes labels.

- Look out for graphs with two Y axes – these need extra special concentration and anyone can make a mistake!

- If you are given space for a calculation you will very likely need to use it! A calculator is essential.

- The main types of calculation tend to be **ratios**, **averages** and **percentages** – make sure you can do these common calculations.

- Answers to calculations will not usually have more than two decimal places.

- Do not leave blanks. Always have a go, using the language in the question if you can.

Good luck!

Remember that the rewards for passing National 5 Biology are well worth it! Your pass will help you get the future you want for yourself. In the exam, be confident in your own ability. If you're not sure how to answer a question, trust your instincts and just give it a go anyway. Keep calm and don't panic! GOOD LUCK!

NATIONAL 5

2013 Specimen
Question Paper

National Qualifications SPECIMEN ONLY

SQ03/N5/01

Biology
Section 1—Questions

Date — Not applicable

Duration — 2 hours

Instructions for completion of Section 1 are given on Page two of the question paper SQ03/N5/02.

Record your answers on the grid on Page three of your answer booklet.

Do NOT write in this booklet.

Before leaving the examination room you must give your answer booklet to the Invigilator. If you do not, you may lose all the marks for this paper.

SECTION 1

1. The diagram below shows one of the stages of mitosis in the root tip of a plant.

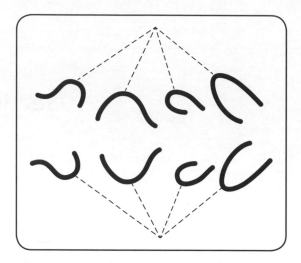

Which of the following statements describes the stage shown?

A   Chromosomes line up at the equator of the cell

B   Daughter chromosomes gather at the ends of the cell

C   Chromosomes become visible as pairs of identical chromatids

D   Spindle fibres pull chromatids to opposite poles of the cell

2. A reaction takes place because the active site of an enzyme is complementary to

A   one type of substrate molecule

B   all types of substrate molecule

C   one type of product molecule

D   all types of product molecules.

3. The diagram below shows stages in the production of a substance, such as insulin, by genetic engineering.

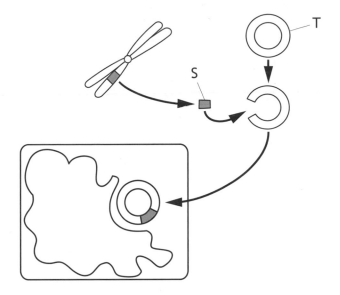

Which line in the table below correctly identifies **S** and **T**?

	S	T
A	Gene	Plasmid
B	Gene	Bacterium
C	Chromosome	Plasmid
D	Chromosome	Bacterium

4. Which of the following shows the correct location and number of ATP molecules released from a molecule of glucose during fermentation?

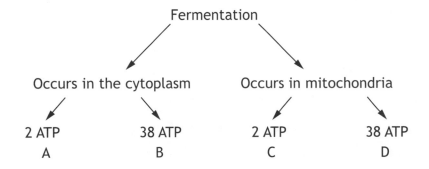

5. The following statements relate to meristems.

1	They produce non-specialised cells
2	They are the sites of gamete production
3	They are found only in plants
4	They are found only in animals

Which of the above statements are correct?

A   1 and 2 only

B   1 and 3 only

C   2 and 3 only

D   2 and 4 only

6. The diagram below shows the movement of food along the oesophagus by peristalsis.

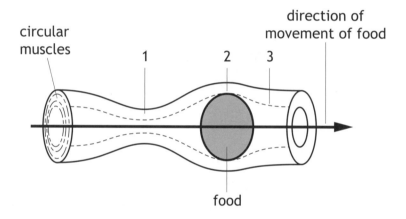

Which line in the table below correctly describes the state of the circular muscles at points 1, 2 and 3 on the diagram?

	Circular muscles		
	Point 1	Point 2	Point 3
A	contracted	relaxed	contracted
B	relaxed	contracted	contracted
C	contracted	relaxed	relaxed
D	relaxed	contracted	relaxed

7. In humans the inheritance of earlobe type is an example of discrete variation. The allele for free earlobes (E) is dominant to the allele for fixed earlobes (e). The diagram below shows the inheritance of this characteristic.

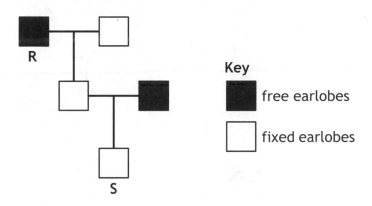

Key

■ free earlobes

□ fixed earlobes

Which line in the table below correctly identifies the genotypes of individuals **R** and **S**?

	Genotype	
	R	**S**
A	EE	ee
B	Ee	ee
C	Ee	Ee
D	ee	EE

8. The diagram below shows some of the structures involved in transport in plants.

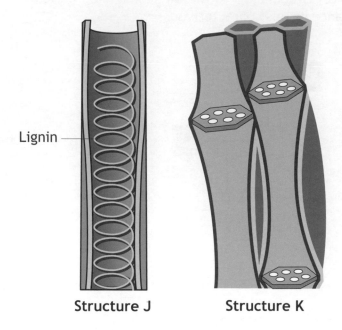

Structure J   Structure K

Which line in the table below correctly identifies structures J and K and the substances transported by them?

	Structure J		Structure K	
	Name	Substance transported	Name	Substance transported
A	Xylem	Water	Phloem	Sugar
B	Xylem	Sugar	Phloem	Water
C	Phloem	Water	Xylem	Sugar
D	Phloem	Sugar	Xylem	Water

9. Transpiration occurs from the leaves of a plant.

Which environmental conditions would produce the greatest transpiration rate?

A   Warm and still air

B   Cold and still air

C   Warm and windy

D   Cold and windy

Questions **10** and **11** refer to the diagram of the heart below.

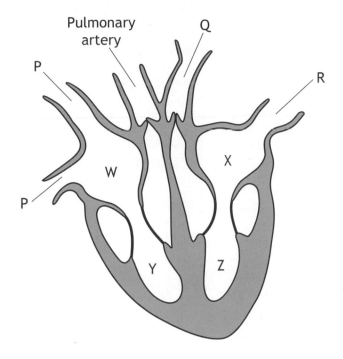

10. Which line in the table below correctly identifies the four chambers of the heart labelled W, X, Y and Z?

	W	X	Y	Z
A	Right ventricle	Left ventricle	Right atrium	Left atrium
B	Right ventricle	Left ventricle	Left atrium	Right atrium
C	Right atrium	Left atrium	Left ventricle	Right ventricle
D	Right atrium	Left atrium	Right ventricle	Left ventricle

11. Which line in the table below correctly identifies the type of blood carried in blood vessels P, Q and R?

	P	Q	R
A	deoxygenated	oxygenated	oxygenated
B	deoxygenated	oxygenated	deoxygenated
C	oxygenated	deoxygenated	oxygenated
D	oxygenated	deoxygenated	deoxygenated

12.　The graph below shows the risk of heart disease and diabetes relative to a body mass index of 21.

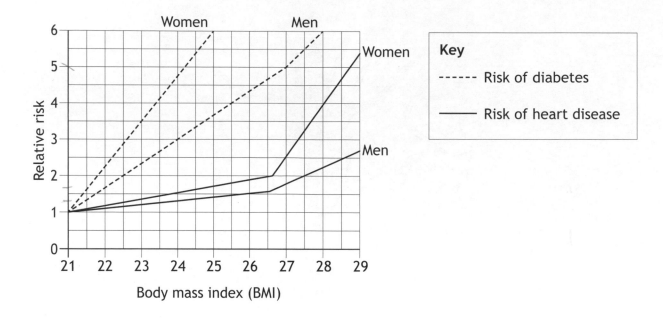

Which of the following statements is correct?

Compared to a BMI of 21

A　women with a BMI of 25 have a six times greater risk of getting heart disease

B　men with a BMI of 24 have a three times greater risk of getting heart disease

C　women with a BMI of 28 have a four times greater risk of getting diabetes

D　men with a BMI of 27 have a five times greater risk of getting diabetes.

13.　The total variety of all living things on Earth is described as

A　an ecosystem

B　biodiversity

C　a population

D　competition.

14.　Which of the following statements about a woodland describes a community?

A　All the oak trees

B　All the plants

C　All the oak trees and blackbirds

D　All the plants and animals

15. The diagram below shows part of a food web in an oak woodland.

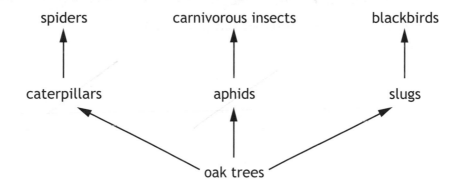

The use of insecticides in a nearby field resulted in the death of most aphids and caterpillars.

Which line in the table identifies the effect on the numbers of slugs and carnivorous insects?

	Number of slugs	*Number of carnivorous insects*
A	decreases	stays the same
B	increases	decreases
C	decreases	increases
D	increases	stays the same

Questions **16** and **17** refer to the following information.

An investigation was carried out into the effect of a hedge on the growth of wheat plants.

Groups of 100 wheat plants were planted at different distances from the hedge. The heights of the plants were measured after six weeks and the results are shown in the table.

Distance planted from hedge (m)	Average height of wheat after six weeks (cm)
2·0	45
2·5	54
3·0	60
3·5	69
4·0	78
4·5	90

16. The reliability of the results was increased by

A   measuring the height of plants after six weeks

B   planting groups of 100 wheat plants

C   planting the wheat plants at different distances from the hedge

D   calculating an average height of wheat plants.

17. What is the percentage increase in average height of wheat planted between 2·0 m and 4·5 m from the hedge?

A   45%

B   50%

C   66%

D   100%

18. One of the roles of decomposers in an ecosystem is to

A   convert protein and waste into ammonia and nitrates

B   produce animal and plant protein from nitrates

C   convert nitrogen from the air into ammonia and nitrates

D   release nitrogen into the air from nitrates.

19. Which one of the following graphs shows the effects of competition for the same food between a successful species and an unsuccessful species?

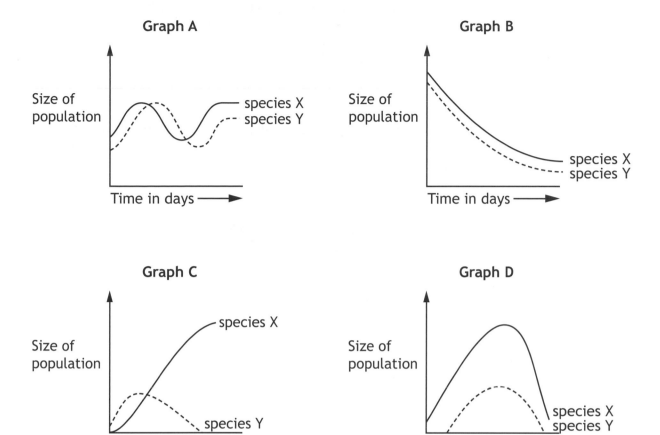

Graph A

Size of population

species X
species Y

Time in days ——▶

Graph B

Size of population

species X
species Y

Time in days ——▶

Graph C

Size of population

species X

species Y

Time in days ——▶

Graph D

Size of population

species X
species Y

Time in days ——▶

20. The diagram below represents a population of animals.

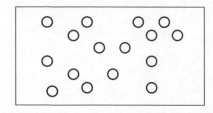

The following diagrams show the stages of speciation occurring from this population.

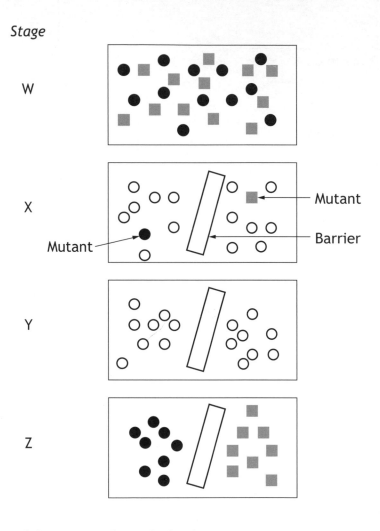

The correct order of the stages of speciation is

A   Z, W, X, Y

B   Z, X, W, Y

C   Y, X, Z, W

D   Y, Z, X, W.

**[END OF SECTION 1. NOW ATTEMPT THE QUESTIONS IN SECTION 2
OF YOUR QUESTION AND ANSWER BOOKLET]**

N5

National
Qualifications
SPECIMEN ONLY

Mark

SQ03/N5/02

Biology
Section 1—Answer Grid
and Section 2

Date — Not applicable

Duration — 2 hours

Fill in these boxes and read what is printed below.

Full name of centre

Town

Forename(s)

Surname

Number of seat

Date of birth

Day	Month	Year
D D	M M	Y Y

Scottish candidate number

Total marks — 80

SECTION 1 — 20 marks

Attempt ALL questions in this section.

Instructions for completion of Section 1 are given on Page two.

SECTION 2 — 60 marks

Attempt ALL questions in this section.

Read all questions carefully before attempting.

Use **blue** or **black** ink. Do NOT use gel pens.

Write your answers in the spaces provided. Additional space for answers and rough work is provided at the end of this booklet. If you use this space, write clearly the number of the question you are attempting. Any rough work must be written in this booklet. You should score through your rough work when you have written your fair copy.

Before leaving the examination room you must give this booklet to the Invigilator. If you do not, you may lose all the marks for this paper.

SQA
©

SECTION 1— 20 marks

The questions for Section 1 are contained in the booklet Biology Section 1—Questions.
Read these and record your answers on the grid on Page three opposite.

1.   The answer to each question is **either** A, B, C or D. Decide what your answer is, then fill in the appropriate bubble (see sample question below).

2.   There is **only one correct** answer to each question.

3.   Any rough working should be done on the additional space for rough working and answers.

Sample Question

The thigh bone is called the

     A   humerus

     B   femur

     C   tibia

     D   fibula.

The correct answer is **B**—femur. The answer **B** bubble has been clearly filled in (see below).

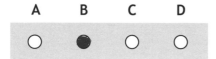

Changing an answer

If you decide to change your answer, cancel your first answer by putting a cross through it (see below) and fill in the answer you want. The answer below has been changed to **D**.

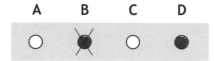

If you then decide to change back to an answer you have already scored out, put a tick (✓) to the **right** of the answer you want, as shown below:

or

SECTION 1 Answer Grid

	A	B	C	D
1	○	○	○	○
2	○	○	○	○
3	○	○	○	○
4	○	○	○	○
5	○	○	○	○
6	○	○	○	○
7	○	○	○	○
8	○	○	○	○
9	○	○	○	○
10	○	○	○	○
11	○	○	○	○
12	○	○	○	○
13	○	○	○	○
14	○	○	○	○
15	○	○	○	○
16	○	○	○	○
17	○	○	○	○
18	○	○	○	○
19	○	○	○	○
20	○	○	○	○

[BLANK PAGE]

SECTION 2 — 60 marks

Attempt ALL questions

1. A variegated leaf contains green areas and white areas.

    A student investigated cells from both areas.

    One of these cells is shown below.

    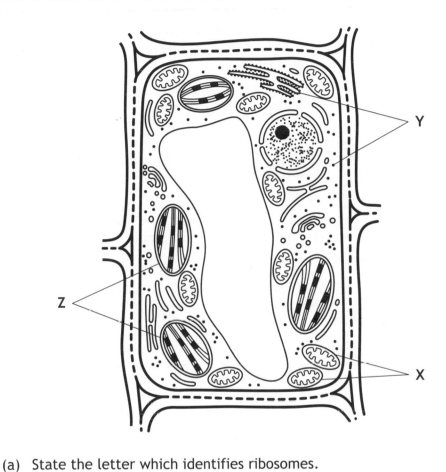

    (a) State the letter which identifies ribosomes.

    _____

    1

    (b) What evidence in the diagram suggests that this cell produces large quantities of ATP?

    _____

    1

    (c) The student concluded that this cell is from the green area. Explain why the student's conclusion is correct.

    _____

    _____

    2

    Total marks 4

2. The diagram below shows a site of gas exchange in the lungs.

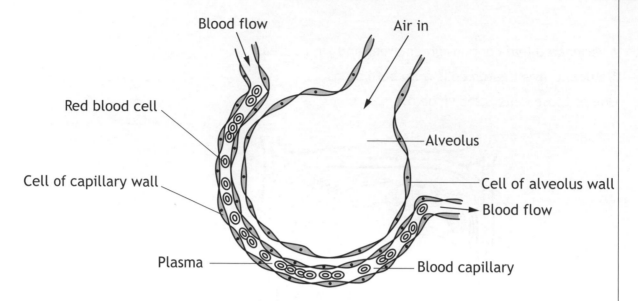

The table below shows the relative concentration of oxygen, carbon dioxide and water in these cells and plasma, the liquid part of the blood.

	Relative concentration of substances		
	oxygen	carbon dioxide	water
Plasma	low	high	medium
Red blood cell	low	high	medium
Cell of capillary wall	medium	medium	medium
Cell of alveolus wall	high	low	medium

(a) (i) Describe the pathway that oxygen would take when moving between these cells.   **1**

_____

_____

(ii) Explain why the oxygen moves along this pathway.   **1**

_____

_____

MARKS | DO NOT WRITE IN THIS MARGIN

2. (continued)

(b) State whether osmosis would occur between the cells of the capillary wall and the cells of the alveolus wall. Insert a tick (✓) in the correct box. **1**

Osmosis would occur ☐ Osmosis would not occur ☐

Justify your answer.

____

____

Total marks 3

MARKS | DO NOT WRITE IN THIS MARGIN

3. The diagram below shows how genetic information in the nucleus is used in the first stage of making a protein.

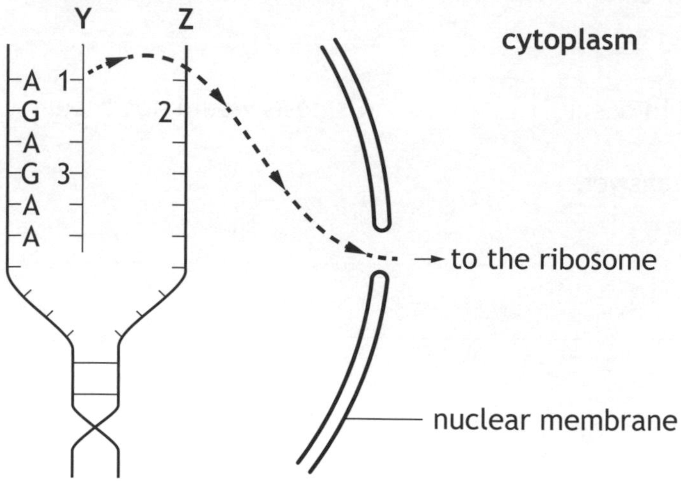

(a) (i) Name molecule **Y**.   **1**

_____

    (ii) Underline one option in each bracket to make the following sentences correct.   **2**

    1. The molecules represented by the letter **A** are ⎰ bases / genes / proteins ⎱

    2. The complementary strand **Z** would have the letter ⎰ A / C / G / T ⎱ at position **2** in the diagram.

(b) Name the basic units which are joined together to make a protein at the ribosome.   **1**

_____

(c) The diagram above shows a section of the code to make a protein such as amylase. Describe how the code to make the protein insulin would differ from this.   **1**

_____

_____

_____

                                            Total marks   5

MARKS

DO NOT WRITE IN THIS MARGIN

4. (a) Photosynthesis is the process by which plants produce sugar using light.

    The flow diagram represents stages of photosynthesis in a leaf.

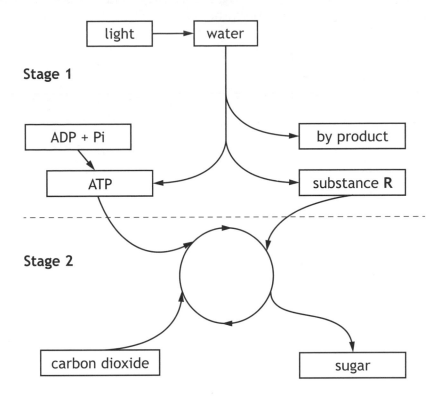

    (i) Identify substance **R**.

    1

    _____

    (ii) Describe the transfer of energy from light arriving at the leaf to the formation of sugar.

    3

    _____

    _____

    _____

    _____

    _____

    _____

DO NOT
WRITE IN
THIS
MARGIN

4. **(continued)**

(b) The graph shows the effect of light intensity and carbon dioxide concentration on the rate of photosynthesis.

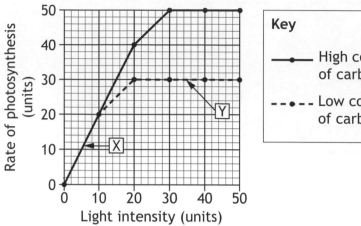

Identify the limiting factor at each of the points X and Y.   1

X _____

Y _____

Total marks   **5**

MARKS | DO NOT WRITE IN THIS MARGIN

5. An investigation was carried out to find the effect of pH on fermentation by yeast, using the apparatus shown. Six groups of students carried out the investigation.

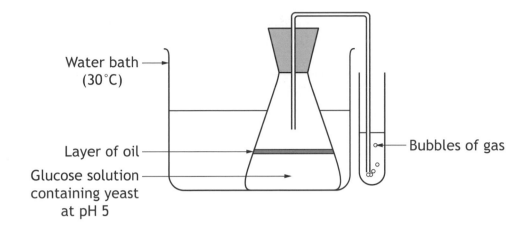

Water bath (30°C)

Layer of oil

Glucose solution containing yeast at pH 5

Bubbles of gas

The investigation was repeated at pH 3, pH 7 and pH 9.

The number of bubbles produced each minute was counted.

Each group carried out the investigation several times and calculated average values for their results, as shown in the table below.

| Group | Average number of bubbles produced per minute | | | |
	pH 3	pH 5	pH 7	pH 9
1	8	25	17	0
2	10	21	13	3
3	15	23	14	0
4	17	22	16	0
5	19	24	12	1
6	22	17	18	9

(a) Name the gas produced during fermentation in yeast.   1

_____

(b) From the table, identify the optimum pH for fermentation by yeast and give a reason for your choice.

pH_____   1

Reason _____   1

MARKS

5. (continued)

(c) This investigation could be adapted to find the effect of a variable other than pH.

Choose **one** variable from the list. Describe **two** ways that the apparatus would be adapted to demonstrate the effect of this variable.   **2**

List

Type of yeast

Temperature

Concentration of glucose solution

Chosen variable _____

Adaptation 1   _____

_____

Adaptation 2   _____

_____

Total marks   5

MARKS | DO NOT WRITE IN THIS MARGIN

6. (a) The diagram shows a hormone, such as insulin, binding with its target cell.

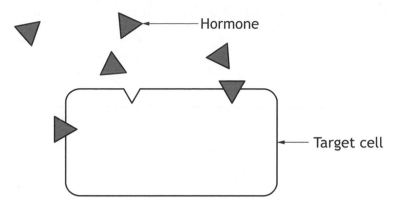

(i) Explain why a hormone only works on its target cell.   1

_____

_____

(ii) Hormone messages travel slower than nerve messages.

State one other difference between these messages.   1

_____

_____

(b) Diabetes is a condition in which the blood glucose level is not fully controlled by insulin. There are two types of diabetes. The table shows information about both types.

Type 1 diabetes	Type 2 diabetes
Insulin is not produced	Insulin is produced but is not used effectively
Usually starts at a young age	Often associated with being obese
Can be triggered by infection	Can be controlled with diet and exercise
Daily insulin injections	Medication can be given in tablet form

MARKS | DO NOT WRITE IN THIS MARGIN

6. (b) (continued)

A person with diabetes was treated with daily insulin injections.

(i) Using information from the table, state which type of diabetes this person had **and** why this treatment was required.

1

_____

_____

(ii) Describe what would happen to this person's blood glucose level if they had not been treated.

1

_____

_____

(iii) Name the organ which is not functioning properly, causing type 1 diabetes.

1

_____

Total marks   5

MARKS

7. The contraceptive pill contains hormones and its use has resulted in small quantities of these hormones reaching our fresh water supplies.

The effect of these hormones on the heart rate of water fleas was investigated. Water fleas were placed into solutions of different hormone concentration and their heart rates were measured.

Hormone concentration (ppm)	0·25	0·50	0·75	1·00	1·25	1·50	1·75	2·00
Average heart rate (beats per second)	4·2	4·2	4·2	3·2	2·9	2·7	2·6	2·4

(a) On the grid below, complete the vertical axis and plot a line graph to show the effect of the hormone on the heart rate of the water fleas.

2

(A spare grid, if required, can be found on *Page twenty-six*)

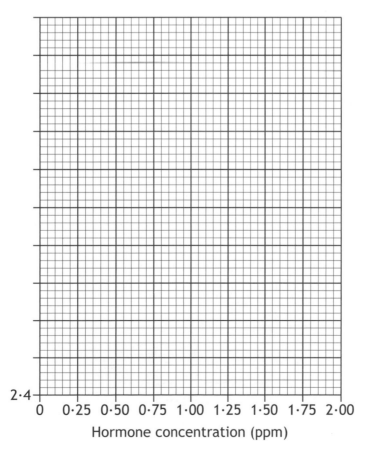

2·4

0   0·25 0·50 0·75 1·00 1·25 1·50 1·75 2·00

Hormone concentration (ppm)

MARKS | DO NOT WRITE IN THIS MARGIN

7.   (continued)

(b)   (i)   Describe the effect of increasing hormone concentration on the heart rate of water fleas.

2

_____

_____

(ii)   It has been suggested that the presence of this hormone in drinking water may have an effect on the heart rate of humans.

State whether you agree or disagree with this suggestion and give a reason to support your choice.

1

Agree ☐   Disagree ☐

Reason _____

_____

_____

(c)   An improvement to this experiment would be to set up a control to count the heart rate of water fleas placed in water without the hormone.

State the purpose of this control.

1

_____

Total marks   6

MARKS

8. Hair type in humans is controlled by a single gene. The dominant form is curly hair (H). The recessive form (h) produces straight hair.

Both parents of this curly-haired child have the genotype Hh.

(a) What term is used to describe the genotype of both parents?   1

_____

(b) Complete the Punnet square to show the possible genotypes of their offspring.   1

Male gametes

		H	h
Female gametes	H		
	h		

(c) State the possible genotype(s) of the girl in the picture.   1

_____

Total marks   3

MARKS | DO NOT WRITE IN THIS MARGIN

9. (a) Blood travels in three types of blood vessels.

Compare the structure of **two** of these types of vessels.   3

_____

_____

_____

_____

_____

_____

(b) Haemoglobin is found in red blood cells. State its function.   1

_____

(c) The diagrams below contain information about the causes of death in two countries in 2010.

Country A   Country B

(i) A student compared the data for heart disease for countries A and B and concluded that country B has a healthier lifestyle.

Explain why this conclusion is incorrect.   1

_____

_____

(ii) Both countries have similar incidence rates for lung diseases.

Describe **one** lifestyle change which someone could make to help reduce their chance of lung disease.   1

_____

_____

Total marks   6

MARKS | DO NOT WRITE IN THIS MARGIN

10. (a) The Cichlid fish below are all found in Lake Malawi in Africa.

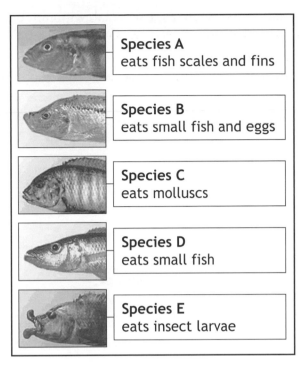

Species A
eats fish scales and fins

Species B
eats small fish and eggs

Species C
eats molluscs

Species D
eats small fish

Species E
eats insect larvae

(i) Using the information shown, identify the feature which enables the fish to have different diets.

1

_____

(ii) Predict **two** species of Cichlid which would be in competition with each other if there was a shortage of fish eggs. Give a reason for your answer.

1

Species _____ and _____

Reason _____

_____

(b) State the term which describes the role that an organism, such as the Cichlid, plays within its community.

1

_____

MARKS | DO NOT WRITE IN THIS MARGIN

10.   (continued)

(c) Fresh water environments, such as Lake Malawi, can be affected by human activities. The overuse of fertilisers can impact on the organisms living in these environments.

Rearrange the following statements to show how this might occur.

1. Chemicals leach into water

2. Fish die

3. Overuse of fertilisers

4. Oxygen levels decrease

5. Algal bloom develops

Place the statement number in the correct box.   1

(d) A fresh water environment is an example of an ecosystem.

Describe what is meant by the term ecosystem.   1

_____

_____

Total marks   5

MARKS | DO NOT WRITE IN THIS MARGIN

11. (a) A food chain is shown below along with three pyramids of numbers.

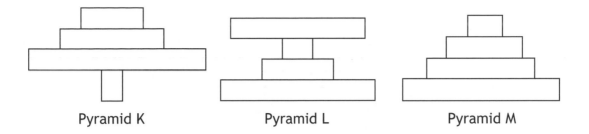

grass ⟶ zebra ⟶ lion ⟶ flea

Pyramid K   Pyramid L   Pyramid M

Identify the pyramid which represents the food chain shown.   **1**

Pyramid _____

(b) This food chain can also be represented by a pyramid of biomass.

State the meaning of the term "pyramid of biomass".   **1**

_____

_____

(c) (i) Calculations were made to estimate the energy content of a food chain involving three species.

heather ⟶ hare ⟶ golden eagle

Two of these values are given in the table below. Complete the table by calculating the missing energy value.   **1**

Space for calculation

Organism	Energy (kJ)
heather	97,000
hare	
golden eagle	970

(ii) State one way in which energy may be lost between stages in a food chain.   **1**

_____

Total marks 4

MARKS | DO NOT WRITE IN THIS MARGIN

12. (a) Root nodules contain bacteria which are involved in the nitrogen cycle.

Bean plants were grown in pots of sand containing different masses of nitrogen fertiliser. After ten weeks, the root nodules were removed, washed and weighed. The average mass of root nodules per plant was calculated.

The results are shown in the graph below.

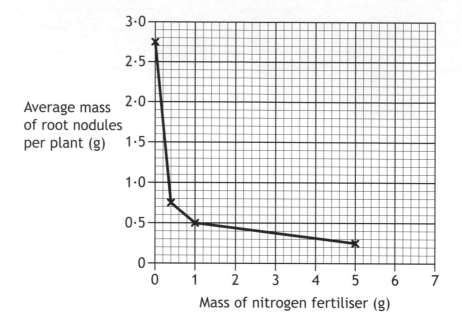

(i) **Using data from the graph**, describe the relationship between the mass of nitrogen fertiliser and the average mass of root nodules per plant.

2

_____

_____

(ii) Predict the average mass of root nodules per plant if 7 g of nitrogen fertiliser were added.

1

_____ g

MARKS | DO NOT WRITE IN THIS MARGIN

12. **(continued)**

(b) The recycling of nitrogen in ecosystems depends on the action of bacteria.

Choose **one** type of bacteria from the list and describe its role in the nitrogen cycle.

1

Types of Bacteria

- Nitrifying bacteria

- Denitrifying bacteria

- Root nodule bacteria

Type of bacteria _____

Role in nitrogen cycle _____

_____

_____

Total marks 4

MARKS

13. (a) **How Europe's bird numbers collapsed**

(*Adapted from The Observer 27/5/12*)

The number of farmland birds in Europe has decreased dramatically in recent years.   A study estimated that the total bird population has dropped from 600 million to 300 million between 1980 and 2009.

It is suggested that changes in farming policies may be largely responsible for this reduction. It has been claimed that intensive farming methods have killed many of the insects eaten by bird species.

The effect on the populations of some bird species is shown in the table below.

Bird species	Population in 1980 (millions)	Population in 2009 (millions)	Population decrease (%)
Linnet	37·0	14·0	62
Meadow pipit	34·9	12·9	63
Corn bunting	27·2	9·2	66
Starling	84·9	39·9	53
Whinchat	10·4	3·4	67
Yellow wagtail	9·4	4·4	53

(i) Explain why the population decrease was expressed as a percentage rather than a decrease in number.

1

_____

_____

(ii) Using information from the passage and the table, calculate the percentage of meadow pipits in the total bird population in 2009.

1

Space for calculation

_____ %

MARKS | DO NOT WRITE IN THIS MARGIN

13. **(a)** **(continued)**

(iii) Which two species of birds were least affected between 1980 and 2009?   1

_____ and _____

(b) One of the advantages of intensive farming is to increase the yield of food crops which can make them more affordable. It also has a number of disadvantages as shown in the list below.

List

- Forests are destroyed to create large open fields and this could lead to soil erosion

- The natural habitats of wild animals are affected

- Loss of biodiversity

- Use of fertilizers can alter the biology of rivers and lochs

- Pesticides sprayed on crops not only destroy pests but also kill beneficial insects

- Toxic effects from pesticides may affect human beings and other organisms when they consume the food crops

Biological control and **GM crops** are methods used in farming.

Choose one of these methods and explain how it is used to overcome some of the disadvantages of intensive farming listed above.   2

Method _____

_____

_____

_____

Total marks   5

[END OF SPECIMEN QUESTION PAPER]

ADDITIONAL SPACE FOR ANSWERS

Spare grid for Question 7 (a)

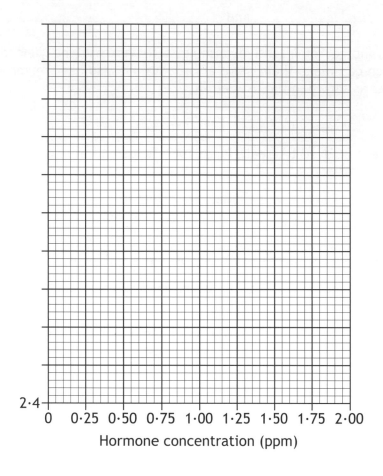

2·4

0　0·25　0·50　0·75　1·00　1·25　1·50　1·75　2·00

Hormone concentration (ppm)

MARKS | DO NOT WRITE IN THIS MARGIN

ADDITIONAL SPACE FOR ROUGH WORKING AND ANSWERS

ADDITIONAL SPACE FOR ROUGH WORKING AND ANSWERS

MARKS

Model Paper 1

Whilst this Model Practice Paper has been specially commissioned by Hodder Gibson for use as practice for the National 5 exams, the key reference documents remain the SQA Specimen Paper 2013 and the SQA Past Paper 2014.

National Qualifications
MODEL PAPER 1

Biology
Section 1—Questions

Duration — 2 hours

Instructions for completion of Section 1 are given on Page two of the question paper.

Record your answers on the grid on Page three of your answer booklet.

Do NOT write in this booklet.

Before leaving the examination room you must give your answer booklet to the Invigilator. If you do not, you may lose all the marks for this paper.

SECTION 1

1. The diagram below shows structures present in a fungal cell.

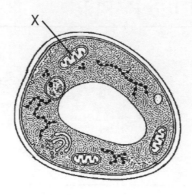

Structure X

A   controls all the cell's activities

B   is the site of protein synthesis

C   is the site of aerobic respiration

D   is the site of photosynthesis.

2. The movement of water from an area of high water concentration to an area of lower water concentration through a selectively permeable membrane is called

A   absorption

B   osmosis

C   plasmolysis

D   active transport.

3. The DNA present in a chromosome carries information that determines the structure and therefore the function of

A   lipids

B   bases

C   carbohydrates

D   proteins.

4. Enzymes act as catalysts because they

    A   are composed of protein

    B   act on any substrate

    C   are unaffected by temperature

    D   speed up chemical reactions in cells.

5. The diagram below shows a stage in the genetic engineering of a bacterium to allow it to produce human insulin.

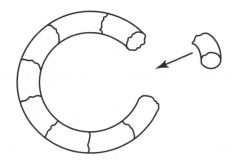

    Which stage of the genetic engineering process is represented in the diagram?

    A   Human insulin gene is inserted into a bacterium

    B   Human insulin gene is inserted into a plasmid

    C   Plasmid is inserted into a human bacterium

    D   Bacterial gene is inserted into a human chromosome

6. The graph below shows the effect of temperature on the rate of photosynthesis in a green plant.

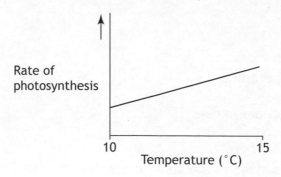

A correct conclusion would be that

A   as the temperature increases, the rate of photosynthesis increases

B   as the temperature increases, the rate of photosynthesis decreases

C   as the temperature increases, the rate of photosynthesis remains constant

D   as the temperature decreases, the rate of photosynthesis increases.

7. The diagram below shows energy transfer within a cell.

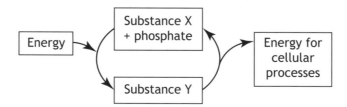

Which line in the table below correctly identifies substances X and Y?

	Substance X	Substance Y
A	ADP	glucose
B	CO_2	ADP
C	ADP	ATP
D	ATP	glucose

8. The list below refers to stages in the response of the nervous system to a stimulus.

1.	Central nervous system sorts information
2.	Electrical impulses sent to muscles
3.	Electrical impulses sent to central nervous system
4.	Sense organ detects the stimulus
5.	Response produced

Which is the correct order of the stages?

A   4, 3, 1, 2, 5

B   3, 4, 2, 1, 5

C   4, 3, 2, 1, 5

D   3, 4, 1, 2, 5

9. The diagram below shows structures present in a flower.

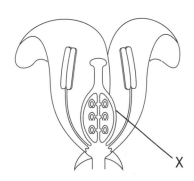

Which line in the table identifies part X and the type of gamete it produces?

	Name of part X	Type of gamete produced
A	ovary	male
B	ovary	female
C	anther	female
D	anther	male

10. Which term refers to the description of a characteristic of an organism?

A   Allele

B   Genotype

C   Phenotype

D   Polygenic

11. The graph below shows the relationship between concentration of oxygen available and the concentration of oxyhaemoglobin in the blood of a mammal.

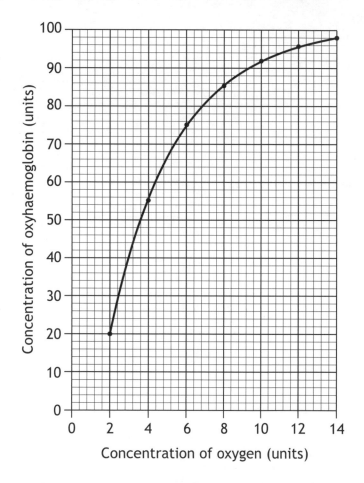

Concentration of oxygen (units)

What is the available increase in the concentration of oxyhaemoglobin when the concentration of available oxygen increases from 2 units to 12 units?

A   10 units

B   26 units

C   76 units

D   380 units

12.   The list below refers to features of the vessels in a capillary network.

1.	Highly branched
2.	In close contact with tissues
3.	Thin-walled

Which statements refer to features that allow efficient gas exchange?

A   1 and 2 only

B   1 and 3 only

C   2 and 3 only

D   1, 2 and 3

13.   The table below shows the changes in the rate of blood circulating in parts of an athlete's body before and during exercise.

Part of body	Rate of blood circulating (cm³ / minute)	
	before exercise	during exercise
Heart muscle	300	900
Skeletal muscles	1200	12000
Skin	600	1900
Muscles of the gut	1500	600

For heart muscle, how many times greater is the rate of blood circulating during exercise compared with before exercise?

A   3 times

B   4 times

C   10 times

D   30 times

14. The table below contains statements referring to the state of muscles which result in food moving through the small intestine.

Statement	State of muscles
1	contracted in front of food
2	relaxed in front of food
3	contracted behind food
4	relaxed behind food

Which statements describe peristalsis?

A   1 and 3

B   1 and 4

C   2 and 3

D   2 and 4

15. Which word describes the organisms living in a habitat and the non-living factors with which they interact?

A   Ecosystem

B   Niche

C   Community

D   Population

16. The diagram below shows a pyramid of biomass.

W represents the total mass of

A   producers

B   prey

C   predators

D   herbivores.

17. Which substances are synthesised by plants using nitrate taken up from soil?

A   Sugars

B   Ammonia

C   Starch

D   Amino acids

18. A survey was carried out to investigate the number of mussels attached to rocks on a sea shore. Quadrats measuring 10cm x 10cm were used in the survey.

The positions of the quadrats and the number of mussels in each quadrat are shown in the diagram below.

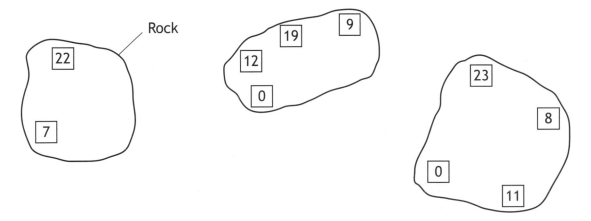

How could the results have been made more reliable?

A   Sample one rock only

B   Use larger quadrats

C   Record a wider variety of species

D   Count each quadrat at the same time of day

19. A sample of fresh soil from woodland was weighed, dried at 60°C and re-weighed. This procedure was repeated until there was no further loss in mass.

The results are shown below.

| Original mass of fresh soil = 50g |
| Final mass of dried soil = 32g |

What percentage of the original soil sample was water?

A   9

B   18

C   36

D   64

20. The Peppered Moth occurs in two different forms. One form is dark coloured and the other light coloured, as shown below.

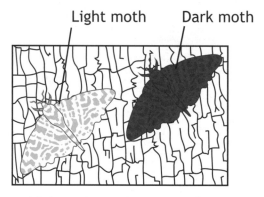

Light moth   Dark moth

Light-coloured tree trunk

The moths are nocturnal and rest by day on the trunks of trees, relying on their camouflage to avoid being seen and eaten by birds.

If the light-coloured tree trunks in an area were darkened by the effects of air pollution, which prediction is most likely to be true?

The numbers of

A   each form would increase

B   the dark form would increase and the light form decrease

C   each form would decrease

D   the light form would increase and the dark form would decrease.

[END OF SECTION 1. NOW ATTEMPT THE QUESTIONS IN SECTION 2 OF YOUR QUESTION AND ANSWER BOOKLET]

National
Qualifications
MODEL PAPER 1

Biology
Section 1—Answer Grid and Section 2

Duration — 2 hours

Total marks — 80

SECTION 1 — 20 marks

Attempt ALL questions in this section.

Instructions for completion of Section 1 are given on Page two.

SECTION 2 — 60 marks

Attempt ALL questions in this section.

Read all questions carefully before attempting.

Use **blue** or **black** ink. Do NOT use gel pens.

Write your answers in the spaces provided. Additional space for answers and rough work is provided at the end of this booklet. If you use this space, write clearly the number of the question you are attempting. Any rough work must be written in this booklet. You should score through your rough work when you have written your fair copy.

Before leaving the examination room you must give this booklet to the Invigilator. If you do not, you may lose all the marks for this paper.

SECTION 1— 20 marks

The questions for Section 1 are contained in the booklet Biology Section 1—Questions.
Read these and record your answers on the grid on Page three opposite.

1. The answer to each question is **either** A, B, C or D. Decide what your answer is, then fill in the appropriate bubble (see sample question below).

2. There is **only one correct** answer to each question.

3. Any rough working should be done on the additional space for rough working and answers.

Sample Question

The thigh bone is called the

     A   humerus

     B   femur

     C   tibia

     D   fibula.

The correct answer is **B**—femur. The answer **B** bubble has been clearly filled in (see below).

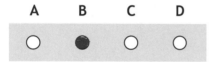

Changing an answer

If you decide to change your answer, cancel your first answer by putting a cross through it (see below) and fill in the answer you want. The answer below has been changed to **D**.

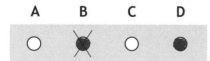

If you then decide to change back to an answer you have already scored out, put a tick (✓) to the **right** of the answer you want, as shown below:

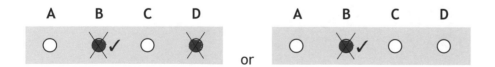

                            or

SECTION 1 Answer Grid

	A	B	C	D
1	○	○	○	○
2	○	○	○	○
3	○	○	○	○
4	○	○	○	○
5	○	○	○	○
6	○	○	○	○
7	○	○	○	○
8	○	○	○	○
9	○	○	○	○
10	○	○	○	○
11	○	○	○	○
12	○	○	○	○
13	○	○	○	○
14	○	○	○	○
15	○	○	○	○
16	○	○	○	○
17	○	○	○	○
18	○	○	○	○
19	○	○	○	○
20	○	○	○	○

[BLANK PAGE]

MARKS | DO NOT WRITE IN THIS MARGIN

SECTION 2 — 60 marks

Attempt ALL questions

1. The diagrams below show two unicellular organisms.

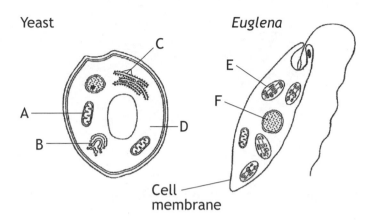

Yeast   *Euglena*

Cell
membrane

(a) (i) Name the two chemical components of the cell membrane.   2

1 _____

2 _____

(ii) Complete the table below by inserting letters from the diagram to show where each process takes place.   2

Process	Letter
mRNA synthesis	
Protein synthesis	
Photosynthesis	

MARKS | DO NOT WRITE IN THIS MARGIN

1. **(continued)**

(b) The diagram below shows a group of human cheek cells as seen under a microscope.

The actual diameter of the field of view was 2 mm.
(1 mm = 1000 micrometres)

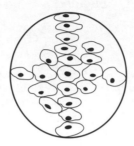

Calculate the average length of the cells in micrometres.
Space for calculation.

Average length _____ micrometres   1

Total marks   5

2. (a) The diagrams below show some stages of cell division in the order in which they occur.

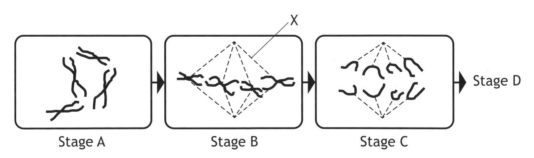

X

Stage A   Stage B   Stage C   Stage D

(i) Identify structure X.   1

_____

MARKS | DO NOT WRITE IN THIS MARGIN

2. **(a) (continued)**

    (ii) Describe what would happen in stage D.   1

_____

_____

(b) The average times taken for the stages of cell division are shown in the table below.

Stage	A	B	C	D
Average time (minutes)	86	34	26	54

Calculate the average time taken for stage B as a percentage of the total time for these four stages.   1

Space for calculation.

_____%

(c) Complete the following sentences by underlining the correct option in each choice bracket.   1

Cells produced by mitosis have $\left\{ \begin{array}{l} \text{one matching set} \\ \text{two matching sets} \end{array} \right\}$ of chromosomes,

and are described as being $\left\{ \begin{array}{l} \text{haploid} \\ \text{diploid} \end{array} \right\}$.

                                     Total marks   4

3. The enzyme catalase breaks down hydrogen peroxide into water and oxygen in living cells.

The apparatus shown below was used to study the effect of temperature on the activity of the enzyme catalase.

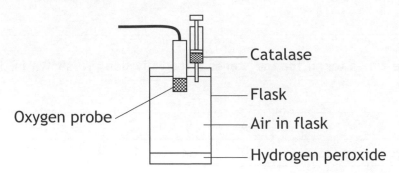

Catalase was added to the flask and the increase in oxygen in the air in the flask was measured by the oxygen probe and recorded as percentage increases.

The procedure was repeated at five different temperatures and the results are shown in the table below.

Temperature (°C)	Increase in oxygen (%)
5	0.5
20	0.8
35	1.4
40	1.1
50	0.1

(a) On the grid below, use the results in the table to complete the line graph of temperature against percentage increase in oxygen.   2
      (A spare grid, if required, can be found on *Page twenty-one.*)

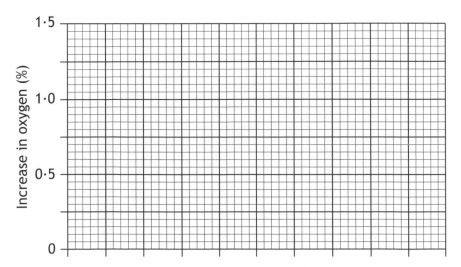

MARKS | DO NOT WRITE IN THIS MARGIN

3. (continued)

(b) Identify the temperature at which catalase was most active? **1**

_____ °C

(c) The catalase and the hydrogen peroxide were both at the required temperature before they were added together.

Explain why this improves the validity of the results. **1**

_____

_____

(d) Explain why no oxygen was produced when the investigation was repeated using a different enzyme. **1**

_____

_____

(e) Describe how the experiment could be modified to investigate the effect of pH on the activity of catalase. **1**

_____

_____

Total marks 6

4. (a) The diagram below shows the two stages of photosynthesis.

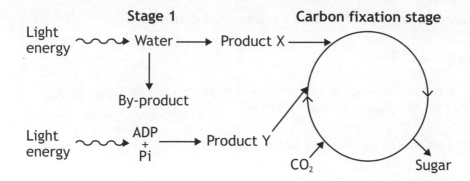

Stage 1   Carbon fixation stage

(i)   Name stage 1.   1

_____

(ii) Identify the by-product released during stage 1.   1

_____

(iii) Name products X and Y produced in stage 1, which are required in
      the carbon fixation stage.   1

      X _____

      Y _____

(iv) Describe the role of carbon dioxide in the carbon fixation stage.   1

_____

_____

MARKS

4. **(continued)**

(b) The graph below shows how light intensity affects the rate of photosynthesis under different conditions.

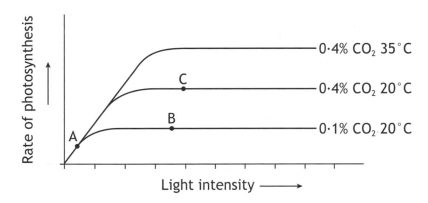

Use the information in the graph to complete the table below.

Tick (✓) **one** box in each row to indicate the factor that is limiting the rate of photosynthesis at points A, B and C.　2

Graph point	Light intensity	CO$_2$ concentration	Temperature
A			
B			
C			

Total marks　6

MARKS | DO NOT WRITE IN THIS MARGIN

5. (a) Stem cells are unspecialised cells in animals which can divide then differentiate to produce a wide range of other cell types in the body.

The chart below shows how some specialised cell types are formed from unspecialised stem cells.

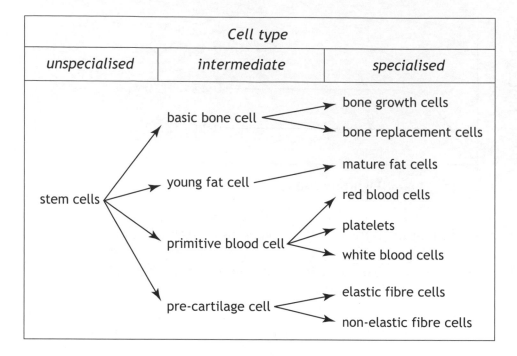

Cell type		
unspecialised	intermediate	specialised

stem cells
- basic bone cell
  - bone growth cells
  - bone replacement cells
- young fat cell
  - mature fat cells
- primitive blood cell
  - red blood cells
  - platelets
  - white blood cells
- pre-cartilage cell
  - elastic fibre cells
  - non-elastic fibre cells

(i) Give the intermediate cell type which can develop into the widest range of specialised cell types.   1

_____

(ii) Give the intermediate cell type which contributes towards growth and repair.   1

_____

(b) (i) Name the sites of production of non-specialised cells in plants.   1

_____

(ii) Name two plant tissues which can be formed when plant cells specialise.   1

1 _____

2 _____

Total marks   4

MARKS | DO NOT WRITE IN THIS MARGIN

6. (a) Different parts of the brain have different functions.

Draw a line from each of the following parts of the brain to its correct function.   **2**

Part	Function
Cerebrum	Controls heart rate
Cerebellum	Controls balance
Medulla	Memory and reasoning

(b) The flowchart below shows structures involved in a reflex arc.

Complete the chart by inserting the names of the missing neurons.   **2**

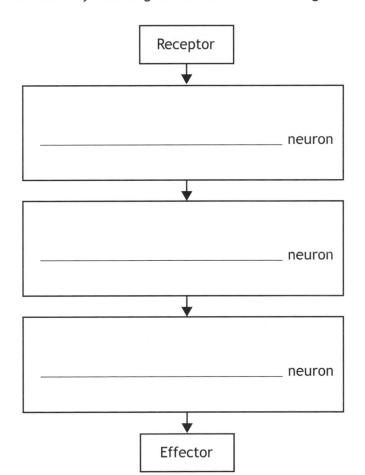

Receptor

_____ neuron

_____ neuron

_____ neuron

Effector

(c) Describe the importance of rapid reflex actions in humans.   **1**

_____

Total marks   5

7. The diagrams below represent human gametes.

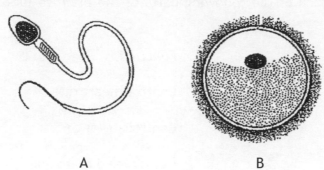

A   B

(a) Name gametes A and B.   1

A _____

B _____

(b) Name the organs that produce gametes A and B.   1

A _____

B _____

(c) Gametes are haploid.

Give the meaning of the term haploid.   1

_____

(d) Name the diploid cell formed as a result of the fusion of gametes.   1

_____

Total marks 4

MARKS | DO NOT WRITE IN THIS MARGIN

8. (a) The diagram below shows cells from a plant root.

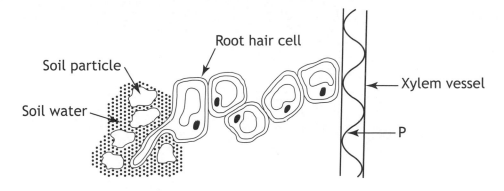

Soil particle

Soil water

Root hair cell

Xylem vessel

P

(i) Describe the uptake of water from soil by plant roots and the role of the root hair cells in this process.   **3**

_____

_____

_____

_____

(ii) Name the strengthening material **P** that lines the xylem vessel.   **1**

_____

(b) Give **one** reason why plants require water.   **1**

_____

Total marks   **5**

MARKS | DO NOT WRITE IN THIS MARGIN

9. The diagram below shows the chambers and blood vessels in a mammalian heart.

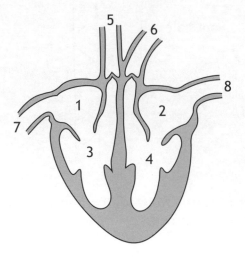

(a) Complete the table below using the correct number from the diagram for each description.   2

Description	Number
Chamber that receives blood from the lungs	
Artery that carries blood from the heart to the body	
Chamber that pumps blood to the lungs	
Vein that carries blood from the body to the heart	

(b) The following sentences give information about blood vessels.

Underline **one** option in each bracket to make the sentence correct.   2

Arteries have a $\left\{\begin{array}{c}\text{thicker}\\\text{thinner}\end{array}\right\}$ muscular wall with a

$\left\{\begin{array}{c}\text{narrower}\\\text{wider}\end{array}\right\}$ internal diameter than veins. Arteries carry

blood at $\left\{\begin{array}{c}\text{low}\\\text{high}\end{array}\right\}$ pressure $\left\{\begin{array}{c}\text{away from}\\\text{back to}\end{array}\right\}$ the heart.

(c) Name the pigment in red blood cells that transports oxygen.   1

_____

Total marks   5

MARKS | DO NOT WRITE IN THIS MARGIN

10. (a) The grid below contains some terms related to the Earth's biomes.

A	B	C	D
Flora	Temperature	Food availability	Fauna

E	F	G	H
Habitat	Rainfall	Population	Predation

Use letters from the grid above to complete the following statements.

(i) Two factors which affect the global distribution of biomes are

_____ and _____.   1

(ii) Biomes are distinguished by their climate, _____and

_____.   1

(iii) An example of a biotic factor is _____.   1

(b) Describe what is meant by the term "biodiversity".   1

_____

_____

(c) Identify **one** example of a human activity which could impact on biodiversity.   1

_____

Total marks   5

11. Six pitfall traps were set in a woodland to sample the invertebrates living there.

    The results are shown in the table below.

Pitfall trap	Number of each type of invertebrate caught				
	spiders	woodlice	beetles	snails	earthworms
1	6	2	2	0	0
2	8	5	5	3	0
3	7	0	3	2	1
4	4	3	7	3	0
5	9	3	0	1	1
6	8	1	4	1	0

(a) Calculate the average number of spiders found per trap.   1

   Space for calculation.

   _____

(b) Express the total numbers of woodlice and beetles trapped as a simple whole number ratio.   1

   Space for calculation.

   _____ : _____
    woodlice   beetles

MARKS | DO NOT WRITE IN THIS MARGIN

11. (continued)

(c) Describe **one** precaution that should be taken when setting up an individual pitfall trap and state why this precaution is necessary.

Precaution   1

_____

_____

Reason   1

_____

_____

(d) Centipedes eat invertebrates.

It was calculated that woodlice make up 10% of their diet and spiders account for 20%. Beetles make up 10% of their diet and other invertebrates make up 60%.

Complete the pie chart below to show this information.   2
(A spare pie chart, if required, can be found on *Page twenty-one*.)

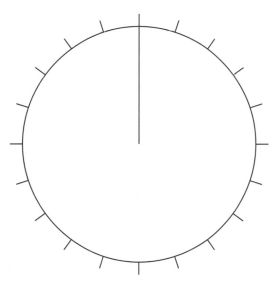

Total marks   6

12. Many species of Cichlid fish are found in Lake Malawi in Africa.

These species have evolved from a single species.

The diagram below shows the heads and mouthparts of three different Cichlid fish species and gives information on their food and feeding methods.

Sucks in microscopic organisms from the water

Scrapes algae from the surfaces of rocks

Crushes snail shells and extracts flesh

(a) Describe the part played by mutations and natural selection in the evolution of these species.   **3**

_____

_____

_____

_____

_____

_____

_____

(b) Describe the advantage each Cichlid species gains from being a specialised feeder.   **1**

_____

(c) Give the term used for the formation of two or more species from a common ancestor.   **1**

_____

Total marks   5

[END OF MODEL PRACTICE PAPER]

MARKS DO NOT WRITE IN THIS MARGIN

ADDITIONAL SPACE FOR ANSWERS

Spare grid for Question 3 (a).

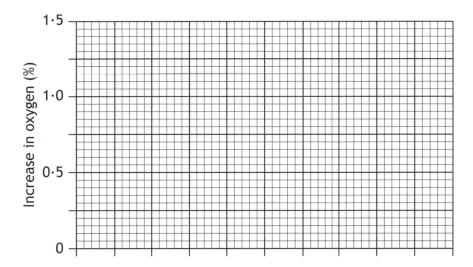

Spare chart for Question 11 (d).

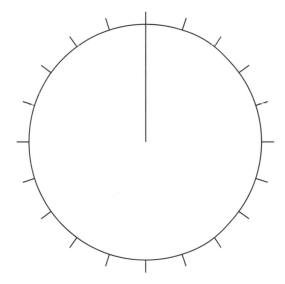

ADDITIONAL SPACE FOR ROUGH WORKING AND ANSWERS

Model Paper 2

Whilst this Model Practice Paper has been specially commissioned by Hodder Gibson for use as practice for the National 5 exams, the key reference documents remain the SQA Specimen Paper 2013 and the SQA Past Paper 2014.

HODDER
GIBSON
LEARN MORE

National Qualifications
MODEL PAPER 2

Biology
Section 1—Questions

Duration — 2 hours

Instructions for completion of Section 1 are given on Page two of the question paper.

Record your answers on the grid on Page three of your answer booklet.

Do NOT write in this booklet.

Before leaving the examination room you must give your answer booklet to the Invigilator. If you do not, you may lose all the marks for this paper.

SECTION 1

1. The diagram below represents a plant cell.

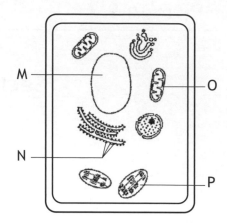

Which parts of the cell would also be found in an animal cell?

A   M and N

B   N and O

C   M and P

D   M, N, O and P

2. A piece of potato was cut and weighed. It was placed in pure water for one hour then removed, dried and re-weighed. Finally, it was placed in a concentrated salt solution for one hour, removed, dried and weighed again.

   Which line in the table below records the results most likely to be obtained by these treatments?

	Original mass of potato (g)	Mass after one hour in pure water (g)	Mass after one hour in concentrated salt solution (g)
A	5	6	4
B	5	4	6
C	6	5	4
D	5	6	6

3. The DNA in a chromosome carries information which determines the order of

    A   proteins

    B   sugars

    C   amino acids

    D   bases.

4. The active site of an enzyme is complementary to

    A   one type of substrate molecule

    B   all types of substrate molecules

    C   one type of product molecule

    D   all types of product molecules.

5. The diagram below shows stages in the genetic engineering of a bacterial cell.

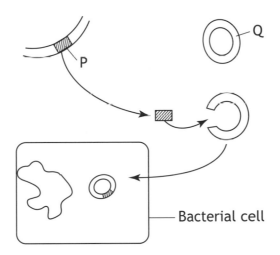

    Which line in the table identifies correctly the structures labelled in the diagram above?

	P	Q
A	plasmid	gene
B	plasmid	chromosome
C	gene	chromosome
D	gene	plasmid

6. The graph below shows the effect of increasing light intensity on the rate of photosynthesis in a green alga.

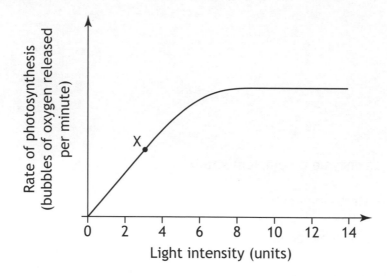

Which factor is limiting the rate of photosynthesis at point X on the graph?

A   Carbon dioxide concentration

B   Temperature

C   Light intensity

D   Oxygen concentration

7. Which of the following substances increases in concentration in contracting human muscle cells when their respiration occurs in the absence of oxygen?

A   Lactic acid

B   Glucose

C   Ethanol

D   ATP

8. The diagram below shows the neurons involved in a reflex arc.

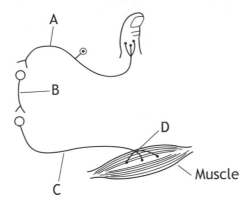

Which letter identifies a relay neuron?

9. The diagram below shows the main parts of a flower.

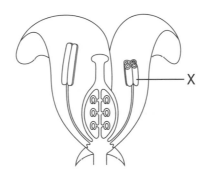

Which line in the table below identifies part X and the type of gamete it produces?

	Name of X	Type of gamete produced
A	anther	female
B	anther	male
C	ovary	female
D	ovary	male

Questions **10** and **11** refer to the diagram below of a human family tree showing the inheritance of tongue rolling.

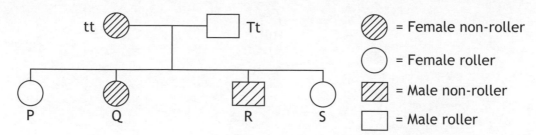

10. The allele for tongue rolling (T) is dominant to the allele for non-rolling (t).

    Which of the following individual(s) in the family tree above are homozygous?

    A   P only

    B   P and Q

    C   P and T

    D   Q and R

11. The chance of individual Q and a non-roller male partner producing a child who is a non-roller is

    A   1 in 1

    B   1 in 2

    C   1 in 3

    D   1 in 4.

12. The diagram below shows the heart and part of the circulatory system in a human.

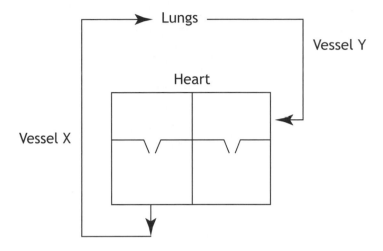

Which line in the table below describes correctly the blood in vessels X and Y?

	Vessel X	Vessel Y
A	deoxygenated	deoxygenated
B	oxygenated	deoxygenated
C	oxygenated	oxygenated
D	deoxygenated	oxygenated

Questions **13** and **14** refer to the diagram below, which shows part of the structure of the breathing system in humans.

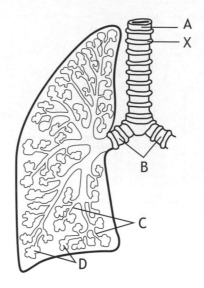

13. Which label identifies alveoli?

14. The function of part X is to

    A   prevent the lungs from collapsing

    B   keep the airway open at all times

    C   prevent food entering the windpipe

    D   trap particles and bacteria.

15. The function of the villi in the small intestine is to increase the surface area for

    A   gas exchange

    B   protection

    C   absorption

    D   peristalsis.

16. A species can be defined as a group of similar organisms which

    A   can breed together to produce fertile offspring

    B   have the same phenotypes

    C   have cells with the same number of chromosomes

    D   contain identical genetic material in their cells.

17. Which of the following describes an ecosystem?

    A   All the members of one species present in a habitat

    B   All the living organisms in a habitat and the non-living components

    C   All the living organisms present in a habitat

    D   All the plants present in a habitat

18. Which of the following describes a niche?

    A   The place where an organism lives

    B   All the living organisms in a habitat and the non-living components

    C   The role of an organism in an ecosystem

    D   All the living organisms present in a habitat

19. Which of the following is an example of intraspecific competition?

    A   Plants of different species competing for different resources

    B   Plants of different species competing for the same resources

    C   Plants of the same species competing for different resources

    D   Plants of the same species competing for the same resources

20. Speciation is considered to have occurred when a population

A   is separated from the rest of the population by an isolation barrier

B   shows increased variation due to the occurrence of new mutations

C   can no longer interbreed with the rest of the population to produce fertile offspring

D   has different selection pressures starting to act upon it.

**[END OF SECTION 1. NOW ATTEMPT THE QUESTIONS IN SECTION 2
OF YOUR QUESTION AND ANSWER BOOKLET]**

National Qualifications
MODEL PAPER 2

Biology
Section 1—Answer Grid and Section 2

Duration — 2 hours

Total marks — 80

SECTION 1 — 20 marks

Attempt ALL questions in this section.

Instructions for completion of Section 1 are given on Page two.

SECTION 2 — 60 marks

Attempt ALL questions in this section.

Read all questions carefully before attempting.

Use **blue** or **black** ink. Do NOT use gel pens.

Write your answers in the spaces provided. Additional space for answers and rough work is provided at the end of this booklet. If you use this space, write clearly the number of the question you are attempting. Any rough work must be written in this booklet. You should score through your rough work when you have written your fair copy.

Before leaving the examination room you must give this booklet to the Invigilator. If you do not, you may lose all the marks for this paper.

SECTION 1— 20 marks

The questions for Section 1 are contained in the booklet Biology Section 1—Questions.
Read these and record your answers on the grid on Page three opposite.

1.   The answer to each question is **either** A, B, C or D. Decide what your answer is, then fill in the appropriate bubble (see sample question below).

2.   There is **only one correct** answer to each question.

3.   Any rough working should be done on the additional space for rough working and answers.

Sample Question

The thigh bone is called the

    A   humerus

    B   femur

    C   tibia

    D   fibula.

The correct answer is **B**—femur. The answer **B** bubble has been clearly filled in (see below).

Changing an answer

If you decide to change your answer, cancel your first answer by putting a cross through it (see below) and fill in the answer you want. The answer below has been changed to **D**.

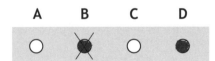

If you then decide to change back to an answer you have already scored out, put a tick (✓) to the **right** of the answer you want, as shown below:

SECTION 1 Answer Grid

	A	B	C	D
1	○	○	○	○
2	○	○	○	○
3	○	○	○	○
4	○	○	○	○
5	○	○	○	○
6	○	○	○	○
7	○	○	○	○
8	○	○	○	○
9	○	○	○	○
10	○	○	○	○
11	○	○	○	○
12	○	○	○	○
13	○	○	○	○
14	○	○	○	○
15	○	○	○	○
16	○	○	○	○
17	○	○	○	○
18	○	○	○	○
19	○	○	○	○
20	○	○	○	○

[BLANK PAGE]

MARKS | DO NOT WRITE IN THIS MARGIN

SECTION 2 — 60 marks

Attempt ALL questions

1. (a) The diagram below represents a bacterial cell.

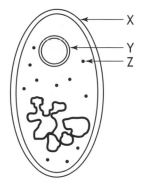

    (i) Name the parts of the cell labelled X and Y.   2

        X _____

        Y _____

    (ii) Give the function of structure Z.   1

        _____

        _____

(b) Give **one** difference and **one** similarity between the structure of a fungal and a bacterial cell.

Difference   1

_____

Similarity   1

_____

Total marks 5

MARKS

2. An investigation was carried out to find the effect of different concentrations of salt solution on the mass of potato tissue.

Five test tubes were set up as shown below, each containing a different concentration of salt solution.

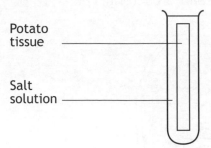

Potato tissue

Salt solution

Each piece of potato tissue was weighed, placed in the solution and left for one hour.

Each was then reweighed and the percentage change in mass was calculated.

The results are shown in the table below.

Salt concentration (g/100cm³)	Change in mass (%)
1	+15
3	+10
6	−5
8	−15
10	−20

MARKS | DO NOT WRITE IN THIS MARGIN

2. (continued)

(a) On the grid below, use the results to complete a line graph of salt concentration against percentage change in mass of the potato tissue. (A spare grid, if required, can be found on *Page twenty-four*.)   **2**

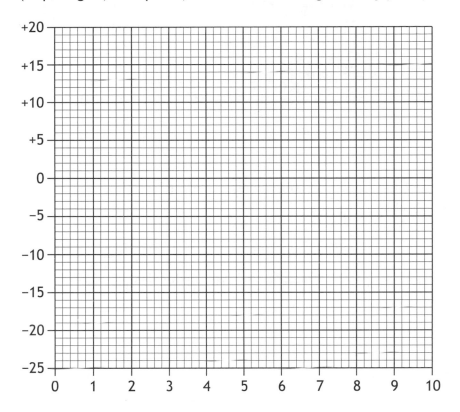

(b) The variety of potato used was the same throughout this investigation.

Give **two** other variables that must be kept constant.   **1**

1 _____

2 _____

(c) Using the results given, predict the salt concentration that has the same concentration as the potato tissue and explain your choice.   **1**

Salt concentration _____ g/100cm^3

Explanation

_____

_____

Total marks   4

3. The diagram below shows features of the ultrastructure of an animal cell.

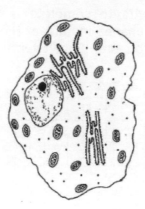

(a) Name the structure, shown in the diagram, which contains the cell's genetic information.

1

_____

(b) The genetic information is encoded in DNA molecules. Describe the structure of a DNA molecule.

3

_____

_____

_____

_____

(c) DNA codes for the amino sequences in protein. Give **two** functions of proteins in cells.

_____

_____

_____

Total marks   6

MARKS | DO NOT WRITE IN THIS MARGIN

4. Genetic engineering can be used to transfer human genes to bacteria artificially.

(a) Complete the boxes below to describe each of the steps carried out to transfer a human gene to a bacterial cell.   **3**

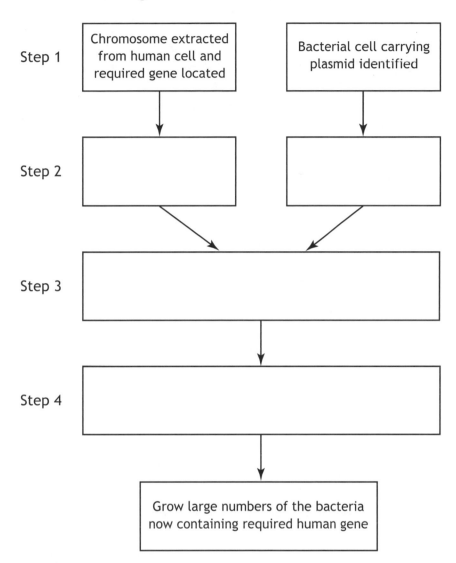

Step 1 — Chromosome extracted from human cell and required gene located | Bacterial cell carrying plasmid identified

Step 2

Step 3

Step 4

Grow large numbers of the bacteria now containing required human gene

(b) Name a human hormone which has been produced by genetically engineered bacteria.

_____   **1**

(c) Describe how genetic information can be transferred naturally between species.

_____   **1**

Total marks   5

5. The diagram below shows stages in the breakdown of glucose in the presence of oxygen to form the final products, gas X and substance Y.

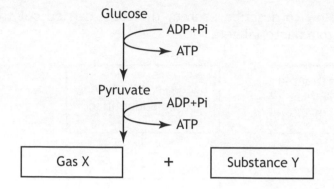

(a) Identify gas X and substance Y.

Gas X   **1**

_____

Substance Y   **1**

_____

(b) State the number of molecules of ATP which are produced per glucose molecule during each of the following pathways.   **1**

Aerobic respiration

_____

Fermentation

_____

(c) State the location of the fermentation pathway in a cell.   **1**

_____

Total marks   **4**

MARKS | DO NOT WRITE IN THIS MARGIN

6. (a) The flowchart below represents part of the control of blood glucose concentration in a human.

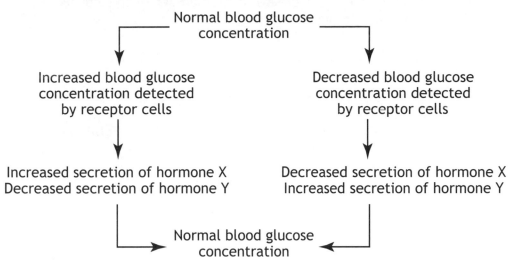

(i) Name the organ which contains the receptor cells that detect changes in blood glucose.   1

_____

(ii) Name hormones X and Y.   2

Hormone X

_____

Hormone Y

_____

(iii) Excess glucose is removed from the blood and stored as carbohydrate in a body organ.   2

Name the storage carbohydrate and the organ in which it is stored.

Storage carbohydrate

_____

Organ

_____

MARKS DO NOT WRITE IN THIS MARGIN

6. (continued)

(b) Give **one** reason for the recent increase in the number of individuals in the Scottish population with diabetes. 1

____

____

Total marks 6

MARKS | DO NOT WRITE IN THIS MARGIN

7. Eye colour in humans shows discrete variation.

The eye colour of 80 school students was recorded and the results are shown in the table below.

Eye colour	Number of school students
Brown	36
Green	12
Blue	24
Grey	4
Hazel	4

(a) Complete the pie chart to show this information.   2
(A spare chart, if required, can be found on *Page twenty-four*.)

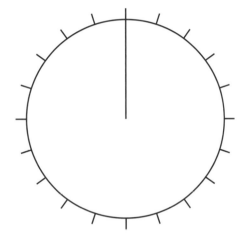

(b) Calculate the percentage of the students with blue eyes.   1

_____ %

MARKS DO NOT WRITE IN THIS MARGIN

7. (continued)

(c) Give the meaning of the terms "continuous variation" and "polygenic inheritance".

Continuous variation **1**

____

____

Polygenic inheritance **1**

____

____

Total marks 5

MARKS | DO NOT WRITE IN THIS MARGIN

8. (a) The graph below shows the rate of water gain and water loss by a plant during a 24-hour period.

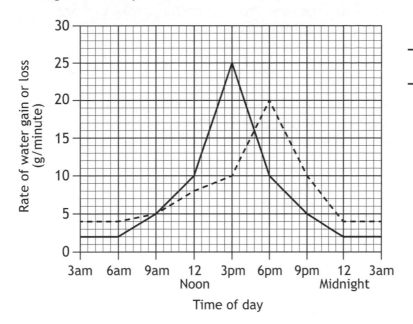

(i)   Calculate the number of hours during which the water loss was greater than 5g per minute.   **1**

Space for calculation.

_____ hours

(ii) Identify the three-hour period during which the greatest increase in the rate of water gain took place.   **1**

Between _____ and _____

(iii) Identify the time in the morning that the rate of water gain equal the rate of water loss?   **1**

_____ am

MARKS | DO NOT WRITE IN THIS MARGIN

8. (continued)

(b) Describe the processes and structures involved in the uptake of water by plant roots and the loss of water through their leaves. **3**

____

____

____

____

____

____

Total marks 6

MARKS | DO NOT WRITE IN THIS MARGIN

9. (a) The diagram below shows an external surface view of a human heart.

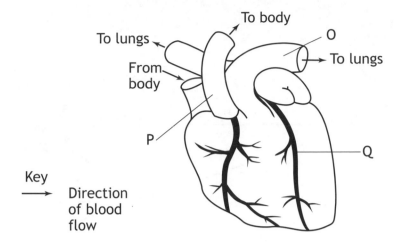

To body

To lungs

From body

O

To lungs

P

Q

Key

⟶   Direction of blood flow

(i) Name the blood vessel O, which carries blood to the lungs.   1

_____

(ii) Name artery P, which carries oxygenated blood to the body.   1

_____

(iii) Name artery Q, which supplies the heart muscle with oxygen.   1

_____

(b) Give **two** examples of lifestyle choices which could **decrease** the chance of a heart attack.

1 _____   1

2 _____   1

Total marks   5

10. The diagram below shows some of the stages in the nitrogen cycle.

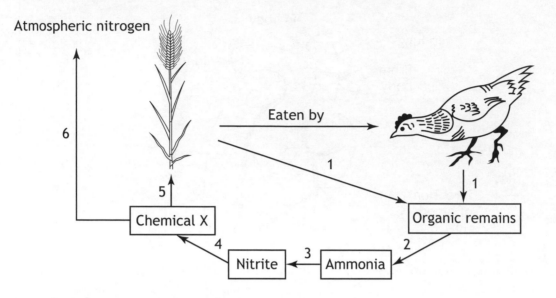

(a) Complete the table below by inserting numbers from the diagram to match each of the named processes. **2**

Process	Number
Uptake	
Decomposition	
Denitrification	
Death	

MARKS | DO NOT WRITE IN THIS MARGIN

10. (continued)

(b) Name chemical X.   1

_____

(c) Name the type of bacteria responsible for stage 3.   1

_____

(d) Describe why nitrogen is needed by living organisms.   1

_____

Total marks   5

11. Boll weevils are insect pests, which feed on cotton plants.

Field trials were carried out on two separate farms to compare the yields of a genetically modified (GM) boll weevil resistant variety of cotton plant with an unmodified variety.

The results are shown in the bar graph below.

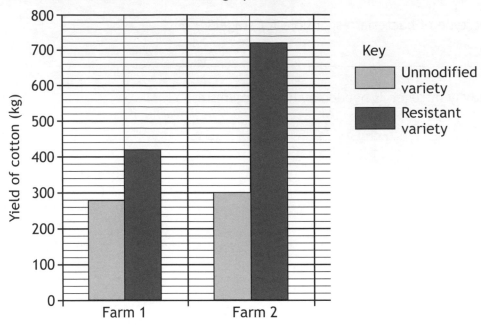

(a) Calculate the average yield of cotton from the unmodified variety from the two farms.   **1**

Space for calculation.

_____ kg

(b) Calculate the percentage increase in yield of cotton from the resistant variety compared to the unmodified variety on Farm 2.   **1**

Space for calculation.

_____ %

MARKS | DO NOT WRITE IN THIS MARGIN

11.   (continued)

(c) The fields planted with the unmodified variety of cotton were used as a control in this field trial.

Give a reason for using this control.   1

_____

_____

(d) Give **one** conclusion that can be drawn from the results.   1

_____

_____

(e) Farmers use pesticides to kill insects, which damage their crops.

Describe an advantage that farmers might gain by planting the GM cotton plant rather than the unmodified variety.   1

_____

_____

Total marks   5

MARKS

12. Lichens can be used to show air pollution caused by sulfur dioxide gas released from burning fossil fuels, such as coal.

The table below gives information about the distribution of lichens in and around a city.

Atmospheric sulfur dioxide (SO_2) levels and the pH of rainwater were also recorded.

Distance from city centre (km)	Number of lichen species (number per km^2)	Atmospheric SO_2 concentration ($\mu g/m^3$)	pH of rainwater
0 – 1.5	0	240	4.6
1.6 – 3.0	1	220	4.8
3.1 – 4.5	7	185	5.0
4.6 – 6.0	13	120	5.5

(a) Describe the relationship between each of the following by underlining one option in each choice bracket.

(i) As the distance from the city centre increases, atmospheric SO_2 levels $\begin{Bmatrix} \text{increase} \\ \text{decrease} \end{Bmatrix}$.

1

(ii) As the atmospheric SO_2 concentration increases, the acidity of the rainwater $\begin{Bmatrix} \text{increases} \\ \text{decreases} \end{Bmatrix}$.

1

(b) From the information in the table above, suggest **one** reason why there were no lichens found within 1.5km of the city centre.

1

_____

_____

MARKS | DO NOT WRITE IN THIS MARGIN

12. **(continued)**

(c) Calculate the average decrease in the atmospheric SO_2 concentration per km over the 6km from the city centre.   1

Space for calculation.

_____ µg/m^3/km

Total marks   **4**

[END OF MODEL PRACTICE PAPER]

ADDITIONAL SPACE FOR ANSWERS

Spare grid for Question 2 (a).

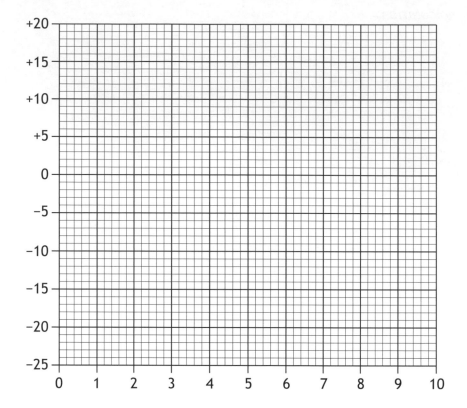

Spare chart for Question 7 (a).

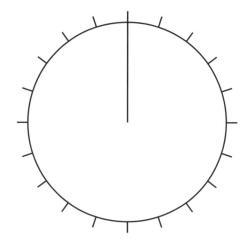

ADDITIONAL SPACE FOR ROUGH WORKING AND ANSWERS

[BLANK PAGE]

Model Paper 3

Whilst this Model Practice Paper has been specially commissioned by Hodder Gibson for use as practice for the National 5 exams, the key reference documents remain the SQA Specimen Paper 2013 and the SQA Past Paper 2014.

Biology
Section 1—Questions

Duration — 2 hours

Instructions for completion of Section 1 are given on Page two of the question paper.

Record your answers on the grid on Page three of your answer booklet.

Do NOT write in this booklet.

Before leaving the examination room you must give your answer booklet to the Invigilator. If you do not, you may lose all the marks for this paper.

SECTION 1

1. Which line in the table below describes correctly the functions of the cell wall and mitochondria in plant cells?

	Function of the cell wall	Function of the mitochondria
A	prevents cell bursting	aerobic respiration
B	controls entry of substances	aerobic respiration
C	prevents cell bursting	photosynthesis
D	controls entry of substances	photosynthesis

2. Which line in the table below summarises correctly the importance of diffusion to an animal cell in terms of raw materials gained and waste products removed?

	Raw material gained	Waste product removed
A	oxygen	glucose
B	carbon dioxide	oxygen
C	oxygen	carbon dioxide
D	glucose	oxygen

3. The genetic code determines the order of

    A   bases in a protein

    B   amino acids in a protein

    C   amino acids in an mRNA strand

    D   sugars in a DNA strand.

4. All enzyme molecules are composed of

    A   carbohydrates

    B   glycerol

    C   proteins

    D   fatty acids.

5. The following steps are involved in the process of genetic engineering.

1	Insertion of a plasmid into a bacterial host cell
2	Use of an enzyme to cut out a piece of chromosome containing a desired gene
3	Insertion of the desired gene into the bacterial plasmid
4	Use of an enzyme to open a bacterial plasmid

Which is the correct sequence of these steps?

A   4, 1, 2, 3

B   2, 4, 3, 1

C   4, 3, 1, 2

D   2, 3, 4, 1

6. The diagram below shows an investigation into factors affecting photosynthesis in a green plant.

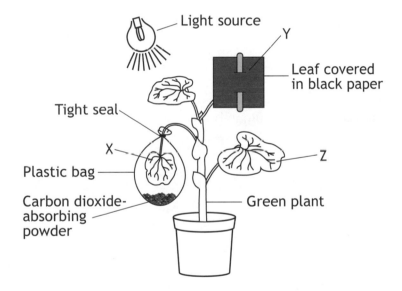

Which of the following statements is true?

A   Leaves X, Y and Z produce sugar

B   Only leaves X and Y produce sugar

C   Only leaf X produces sugar

D   Only leaf Z produces sugar

7. The experiment below was set up to investigate aerobic respiration in an insect.

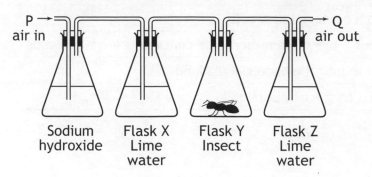

Sodium hydroxide solution absorbs carbon dioxide from air.

Lime water turns from clear to cloudy in the presence of carbon dioxide.

Air is drawn through the apparatus from P to Q, passing through each flask in turn.

If two insects instead of one were placed in flask Y, the lime water in

A   flask X would turn cloudy more slowly

B   flask X would turn cloudy more quickly

C   flask Z would turn cloudy more slowly

D   flask Z would turn cloudy more quickly.

8. The diagram below shows a section through the human brain.

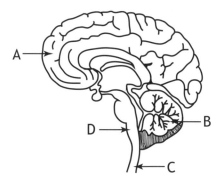

Which letter indicates the site of memory and reasoning?

9. Which line in the table below identifies correctly the effects of a decrease in blood glucose level on the concentrations of insulin and glucagon in the blood.

	Insulin concentration	Glucagon concentration
A	increases	increases
B	increases	decreases
C	decreases	increases
D	decreases	decreases

10. Which line in the table below identifies correctly the male gametes and the site of its production in a flowering plant?

	Male gamete	Site of production
A	sperm	testes
B	pollen nucleus	anther
C	sperm	anther
D	pollen nucleus	testes

11. Inherited characteristics controlled by alleles of more than one gene are described as

    A   homozygous

    B   polygenic

    C   discrete

    D   dominant.

12. Transpiration is the

    A   evaporation of water through stomata

    B   uptake of water by root hair cells

    C   transport of water through xylem

    D   transport of sugars through phloem.

13.   The bar chart below shows the volume of blood supplied to the skeletal muscles and to other parts of the body of a healthy male at rest and during exercise.

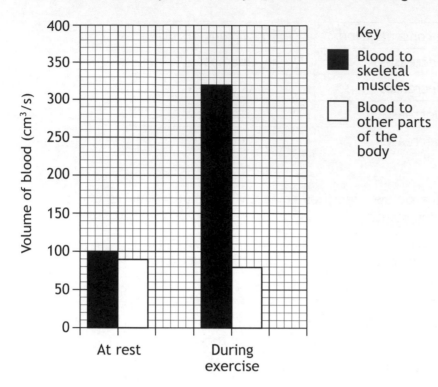

What is the ratio of blood supplied to the skeletal muscles to the blood supplied to other parts of the body during exercise?

A   1:4

B   4:1

C   10:8

D   10:9

14.   The diagram below shows an alveolus with part of its blood capillary network.

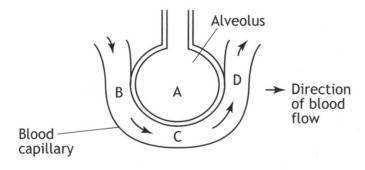

At which position would blood with the highest concentration of oxygen be found?

15. The diagram below shows some structures in a villus.

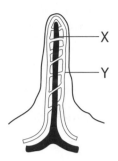

Which line in the table below correctly identifies the product(s) of digestion which pass into structures X and Y?

	Product(s) of digestion passing into X	Product(s) of digestion passing into Y
A	glucose	amino acids
B	glycerol	fatty acids
C	amino acids	glycogen
D	fatty acids	glucose

16. The bar chart below shows the results of a survey into the heights of bell heather plants in an area of Scottish moorland.

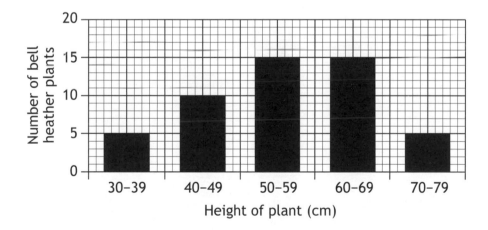

Which percentage of bell heather plants have heights greater than 59cm?

A   20%

B   30%

C   40%

D   70%

17. The information below shows the niche requirements of some woodland bird species.

Bird species	Food	Nest site
Lesser spotted woodpecker	insects	hole excavated in dead wood
Green woodpecker	ants and other insects	hole excavated in live wood
Great spotted woodpecker	insects	hole excavated in live wood
Treecreeper	insects, spiders and other invertebrates	behind loose bark

Between which two bird species will interspecific competition be greatest?

A   Lesser spotted woodpecker and treecreeper

B   Treecreeper and green woodpecker

C   Green woodpecker and great spotted woodpecker

D   Great spotted woodpecker and treecreeper

Questions **18** and **19** refer to the diagram below showing some of the stages in the nitrogen cycle.

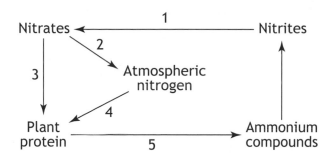

18. Which number represents the action of nitrifying bacteria?

A   1

B   2

C   3

D   4

19. Which number represents the action of decomposers?

A   2

B   3

C   4

D   5

20. Which term refers to the process by which organisms that are better adapted to their environment survive and breed?

A   Natural selection

B   Speciation

C   Mutation

D   Genetic engineering

**[END OF SECTION 1. NOW ATTEMPT THE QUESTIONS IN SECTION 2
OF YOUR QUESTION AND ANSWER BOOKLET]**

[BLANK PAGE]

National
Qualifications
MODEL PAPER 3

Biology
Section 1—Answer Grid
and Section 2

Duration — 2 hours

Total marks — 80

SECTION 1 — 20 marks

Attempt ALL questions in this section.

Instructions for completion of Section 1 are given on Page two.

SECTION 2 — 60 marks

Attempt ALL questions in this section.

Read all questions carefully before attempting.

Use **blue** or **black** ink. Do NOT use gel pens.

Write your answers in the spaces provided. Additional space for answers and rough work is provided at the end of this booklet. If you use this space, write clearly the number of the question you are attempting. Any rough work must be written in this booklet. You should score through your rough work when you have written your fair copy.

Before leaving the examination room you must give this booklet to the Invigilator. If you do not, you may lose all the marks for this paper.

SECTION 1— 20 marks

The questions for Section 1 are contained in the booklet Biology Section 1—Questions.
Read these and record your answers on the grid on Page three opposite.

1. The answer to each question is **either** A, B, C or D. Decide what your answer is, then fill in the appropriate bubble (see sample question below).

2. There is **only one correct** answer to each question.

3. Any rough working should be done on the additional space for rough working and answers.

Sample Question

The thigh bone is called the

    A   humerus

    B   femur

    C   tibia

    D   fibula.

The correct answer is **B**—femur. The answer **B** bubble has been clearly filled in (see below).

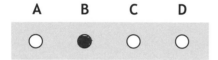

Changing an answer

If you decide to change your answer, cancel your first answer by putting a cross through it (see below) and fill in the answer you want. The answer below has been changed to **D**.

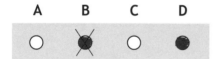

If you then decide to change back to an answer you have already scored out, put a tick (✓) to the **right** of the answer you want, as shown below:

        or

SECTION 1 Answer Grid

	A	B	C	D
1	○	○	○	○
2	○	○	○	○
3	○	○	○	○
4	○	○	○	○
5	○	○	○	○
6	○	○	○	○
7	○	○	○	○
8	○	○	○	○
9	○	○	○	○
10	○	○	○	○
11	○	○	○	○
12	○	○	○	○
13	○	○	○	○
14	○	○	○	○
15	○	○	○	○
16	○	○	○	○
17	○	○	○	○
18	○	○	○	○
19	○	○	○	○
20	○	○	○	○

[BLANK PAGE]

MARKS | DO NOT WRITE IN THIS MARGIN

SECTION 2 — 60 marks

Attempt ALL questions

1. (a) Cells from onion tissue were examined under a microscope and then placed in a concentrated sugar solution for 30 minutes.

   The diagrams below show a cell from the onion before and after being placed in the solution.

   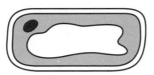   

   Before   After

   (i) Movement of water caused the change in the appearance of the cell.

   Name the process responsible for the movement of water.   1

   _____

   (ii) In terms of water concentration, explain how water movements account for the appearance of the cell after being placed in the solution.   1

   _____

   _____

   (b) Name **one** substance, other than water, which must be able to pass into or out of a cell and explain the importance of the substance to the cell.

   Substance   1

   _____

   Importance   1

   _____

   (c) Give **one** difference between active transport and passive transport.   1

   _____

   _____

   Total marks   5

MARKS

2. The graph below shows the growth curve of a population of bacteria in a fermenter at 30°C over a 24-hour period.

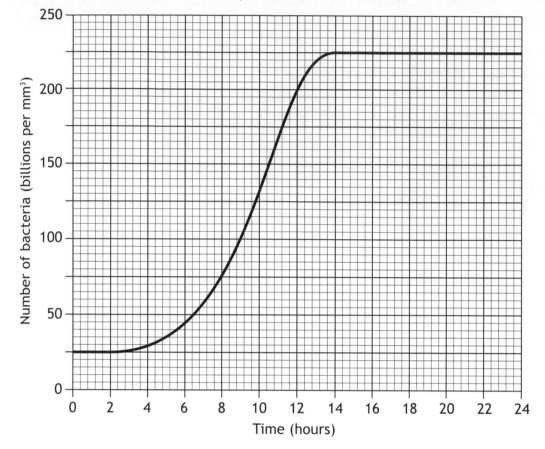

(a) Describe the relationship between the number of bacteria in the population and time.

2

_____

_____

(b) During which two-hour period was there the greatest increase in the number of bacteria?

1

Tick (✓) the correct box

  6–8 hours   8–10 hours   10–12 hours   12–14 hours

MARKS | DO NOT WRITE IN THIS MARGIN

2. **(continued)**

(c) Calculate the percentage increase in the number of bacteria over the first 8 hours.

1

Space for calculation.

_____ %

(d) Express the number of bacteria at 24 hours to the number at the start as a simple whole number ratio.

1

Space for calculation.

_____   _____

at start   :   at 24 hours

(e) Give **one** reason why cell production by cell culture requires aseptic techniques.

1

_____

_____

Total marks   6

3. An investigation was carried out into digestion of a protein.

Protein was mixed with agar gel in a Petri dish. Three wells were cut in the gel. A solution of the enzyme pepsin was placed in one of the wells and solutions of the enzymes amylase and lipase into the remaining wells. The dishes were then left for two days at 20°C.

The experiment was repeated five times.

The diagram below shows the appearance of one of the Petri dishes after two days.

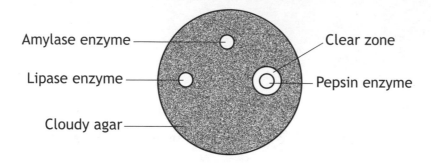

Amylase enzyme — Clear zone
Lipase enzyme —
Pepsin enzyme
Cloudy agar —

The clear zone in the gel around the pepsin well shows that digestion of protein has occurred.

The widths of the clear zones in each dish were measured.

(a) The table below shows the results for each dish.

Petri dish	Width of clear zone around well containing pepsin (mm)
1	4.5
2	4.1
3	4.0
4	4.6
5	3.8
average	

Complete the table by calculating the average width of the clear areas.   1

MARKS

3. **(continued)**

(b) Describe **one** precaution, not already mentioned, that would have to be taken for each dish to ensure the validity of the results.   **1**

_____

(c) Identify the feature of enzyme activity shown by the results.   **1**

_____

(d) State why enzymes can be described as biological catalysts.   **1**

_____

(e) Give **one** condition which could cause an enzyme to become denatured.   **1**

_____

Total marks   5

4. The diagram below shows the two stages of photosynthesis.

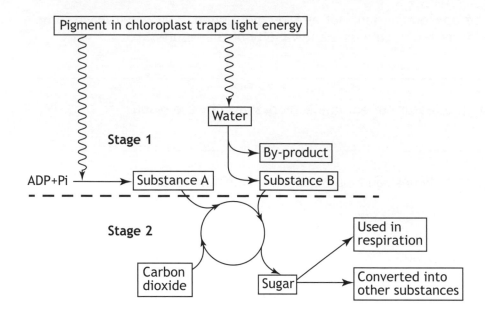

(a) Name the pigment in chloroplasts which traps the light energy required for photosynthesis.

1

_____

(b) In the light-dependent stage of photosynthesis, light energy is converted to chemical energy.

Identify substance A, in which the chemical energy is stored.

1

_____

(c) Identify substance B produced in stage 1, which is required for stage 2.

1

_____

(d) Give **one** substance into which the sugar produced in stage 2 can be converted.

1

_____

(e) Explain why temperature can limit the rate of photosynthesis.

1

_____

_____

Total marks 5

MARKS | DO NOT WRITE IN THIS MARGIN

5. The diagram below shows part of an investigation into the effect of adding two different concentrations of ATP solution to two pieces of muscle tissue.

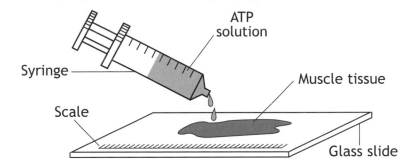

The results of the investigation are given in the table below.

Muscle tissue	Concentration of ATP solution added (%)	Length of muscle tissue (mm)			
		At start	After 5 minutes	Decrease in length	Percentage decrease (%)
1	0.5	48	45.6	2.4	5
2	1.0	45	40.5	4.5	

(a) (i) Calculate the percentage decrease in length of muscle tissue 2.   **1**

Space for calculation.

_____ %

(ii) Give a conclusion that can be drawn from the results.   **1**

_____

_____

(iii) Explain why it is necessary to express the results as a percentage decrease when comparing the results obtained.   **1**

_____

_____

5. **(continued)**

(b) Explain why different syringes should be used to add the ATP solutions in this investigation.

1

_____

(c) The list below contains some features of respiration in germinating peas.

List	
W	Does not require oxygen
X	Releases CO_2
Y	Produces 38 molecules of ATP per glucose molecule
Z	Produces ethanol

Complete the table below by entering the letters from the list in the correct box to match the features with the type of respiration occurring.

2

Each letter may be used once or more than once.

Aerobic respiration in germinating peas	Fermentation in germinating peas

Total marks 6

MARKS

6. The diagram below represents the structures involved in a reflex action which occurs when a pain receptor in a human finger is stimulated.

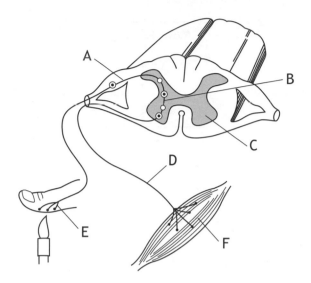

(a) Complete the table below by adding the correct letters from the diagram to identify the stages of the reflex action.   1

Stage	Letter
Stimulus detected by sensory receptor	E
Electrical impulse passes along a sensory neuron	
Electrical impulse sent along a relay neuron	
Electrical impulse passes along a motor neuron	
Effector organ which makes response	

(b) Name the gap between neurons.   1

_____

(c) Describe how information is passed across a gap between two neurons.   1

_____

_____

Total marks   3

7. (a) The flow chart below shows a summary of events that occur during reproduction in a flowering plant.

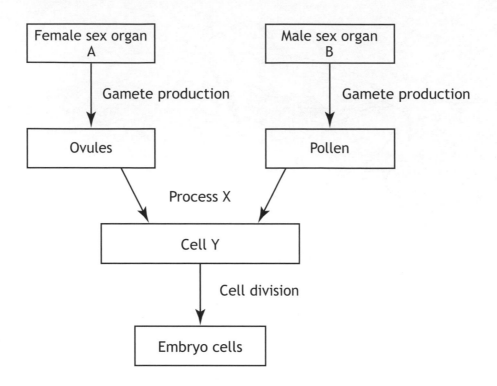

Female sex organ
A

Male sex organ
B

Gamete production

Gamete production

Ovules

Pollen

Process X

Cell Y

Cell division

Embryo cells

(i) Name sex organs A and B.   2

A _____

B _____

(ii) Name process X, which involves the fusion of the gametes.   1

_____

(iii) Name cell Y.   1

_____

MARKS DO NOT WRITE IN THIS MARGIN

7. (continued)

(b) Complete the table below by inserting ticks (✓) into the correct boxes to show which of the cells in the diagram are haploid and which are diploid. **2**

Cell	*Haploid*	*Diploid*
Ovule		
Pollen		
Cell Y		

Total marks 6

8. Tongue rolling is an inherited characteristic in humans.

Tongue rolling is determined by the dominant form of the gene, T and the non-rolling condition is determined by the recessive t.

The family tree diagram below shows the pattern of inheritance in one family.

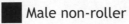

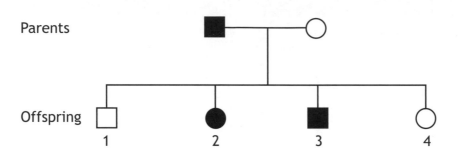

(a) (i) State the genotypes of the following individuals.   2

Male 1

_____

Female 2

_____

Female 4

_____

(ii) Identify which of the parents is homozygous.   1

Tick (✓) the correct box.

Male parent ☐

Female parent ☐

Both parents ☐

Neither parent ☐

(b) Give the term used for different forms of the same gene.   1

_____

MARKS | DO NOT WRITE IN THIS MARGIN

8. (continued)

(c) Tongue rolling is an example of a discrete variation.

Describe what is meant by the term "discrete variation". **1**

____

____

Total marks 5

MARKS

9. (a) The following are types of mammalian blood vessel.

Artery   **Vein**   **Capillary**

Choose **one** of these vessels and describe its structure and function.

Choice   3

_____

_____

_____

_____

_____

(b) Blood is an example of a tissue.

(i) Describe what is meant by the term "tissue".   1

_____

(ii) Following bleeding, lost red blood cells are replaced by the activity of bone marrow.

Identify the type of cell, present in bone marrow, which can divide and differentiate to produce red blood cells.   1

_____

Total marks   5

MARKS | DO NOT WRITE IN THIS MARGIN

10. (a) An insect-resistant variety of crop plant was produced by transferring genes from a bacterial species that produces a toxin that kills insects.

Give the term used to describe a crop plant that has been altered in this way.

1

_____

(b) Biological control methods are sometimes used to control pests that affect crop plants.

Describe what is meant by the biological control of pests.

1

_____

_____

(c) Fertiliser added to farmland can sometimes have an impact on biodiversity in ponds and rivers close by.

Explain this effect.

3

_____

_____

_____

_____

Total marks 5

11. (a) The diagram below shows a stretch of river into which untreated sewage is discharged.

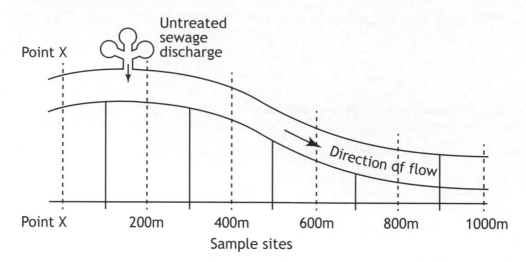

Untreated sewage discharge

Point X

Point X   200m   400m   600m   800m   1000m

Direction of flow

Sample sites

The table below shows the results of a survey carried out to determine the oxygen content of the river water at sample sites above and below the sewage discharge.

Distance of sample site from point X (m)	Oxygen content (units)
0 (at point X)	1.20
200	0.10
400	0.20
600	0.40
800	1.00

11. (a) (continued)

(i) On the grid below, complete the line graph of the distance from point X against the oxygen content of the river.

(A spare grid, if required, can be found on *Page twenty-three*.)

2

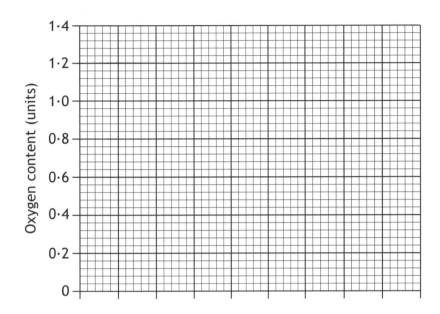

(ii) From the information in the table, calculate how many times greater the oxygen content is at point X compared with the 200m sample site.

1

Space for calculation.

_____ times

(b) Describe how indicator organisms could be sampled to show the effects of the untreated sewage on the river.

2

_____

_____

_____

Total marks 5

12.   Rabbits were introduced to Australia by European settlers.

The graph below shows the change in the rabbit population since their introduction at time A.

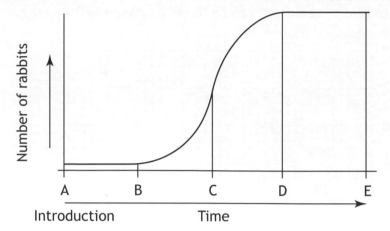

(a)   Describe **one** reason for the change in rabbit population between B and C.   1

_____

_____

(b)   Suggest **one** reason for the population levelling off after point D.   1

_____

(c)   To control over-grazing by rabbits, a disease was introduced at point E. The rabbit population decreased significantly.

Later, the population then started to recover due to a genetic mutation that made some individual rabbits resistant to the disease.

(i)   Describe what is meant by the term "genetic mutation".   1

_____

_____

(ii)   Give the term used to describe the survival of those rabbits with the beneficial mutation, which gave them resistance to the disease.   1

_____

Total marks   4

[END OF MODEL PRACTICE PAPER]

MARKS | DO NOT WRITE IN THIS MARGIN

ADDITIONAL SPACE FOR ANSWERS

Spare grid for Question 11 (a).

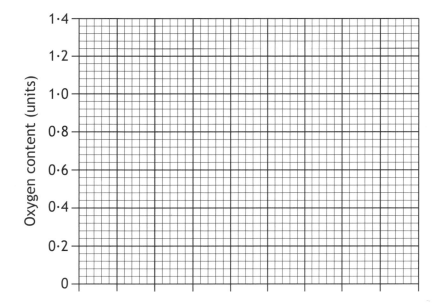

ADDITIONAL SPACE FOR ROUGH WORKING AND ANSWERS

NATIONAL 5

2014

National Qualifications 2014

X707/75/02

Biology
Section 1—Questions

FRIDAY, 16 MAY

9:00 AM – 11:00 AM

Instructions for the completion of Section 1 are given on Page two of your question and answer booklet X707/75/01.

Record your answers on the answer grid on Page three of your question and answer booklet.

Before leaving the examination room you must give your question and answer booklet to the Invigilator; if you do not, you may lose all the marks for this paper.

SECTION 1

1. Which structural feature is found in a plant cell and not in an animal cell?

   A   Nucleus

   B   Cell wall

   C   Cell membrane

   D   Cytoplasm

2. Which line in the table below identifies the direction of diffusion of the three substances during muscle contraction?

	Substance		
	Glucose	Oxygen	Carbon dioxide
A	out	out	in
B	in	out	in
C	out	in	out
D	in	in	out

3. The diagram below represents a genetically engineered bacterial cell.

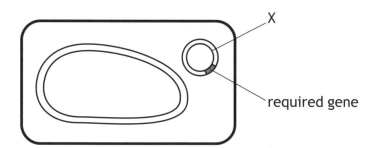

   The structure labelled X is a

   A   chromosome

   B   plasmid

   C   ribosome

   D   nucleus.

4. The light energy for photosynthesis is captured by

    A   water

    B   hydrogen

    C   chlorophyll

    D   oxygen.

5. The diagram below represents the human brain.

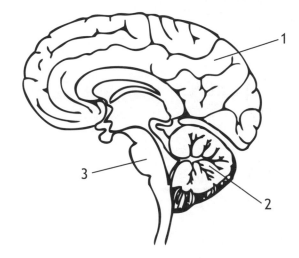

Which line in the table below identifies structures 1, 2 and 3 of the human brain?

	Structure 1	Structure 2	Structure 3
A	medulla	cerebrum	cerebellum
B	cerebrum	medulla	cerebellum
C	cerebellum	cerebrum	medulla
D	cerebrum	cerebellum	medulla

6. Proteins have different functions. Which of the following statements identifies a **protein** and its function?

    A   Hormones carry oxygen around the body.

    B   Enzymes carry chemical messages around the body.

    C   Antibodies defend the body against disease.

    D   Cellulose provides strength and structure to a plant cell wall.

[Turn over

7. Which of the diagrams below identifies neurons and the direction of travel of nerve impulses?

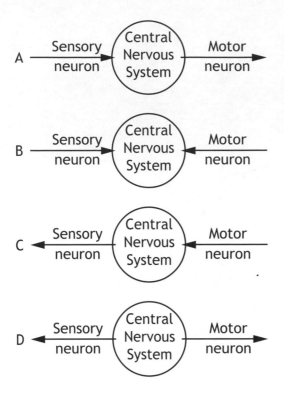

8. Which of the following pairs of human cells have the same number of chromosomes?

    A   Liver cell and sperm cell

    B   Kidney cell and sperm cell

    C   Kidney cell and liver cell

    D   Liver cell and egg cell

9. The table below shows the results of an investigation into the effect of temperature on egg laying in adult red spider mites.

Feature	Temperature (°C)		
	20 °C	25 °C	30 °C
Average length of egg laying period (days)	24	18	12
Average number of eggs laid per female during egg laying period	72	72	72

As the temperature increases, the average number of eggs laid per female per day

A   increases

B   decreases

C   stays the same

D   halves.

10. The following diagrams show a cell at four different stages of mitosis.

1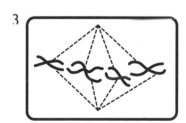

2

3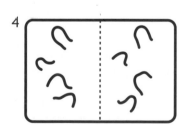

4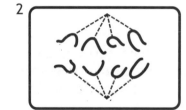

The correct order of the stages of mitosis is

A   1,3,2,4

B   2,3,4,1

C   3,2,1,4

D   4,1,2,3.

11. Which of the following diagrams represents the process of fertilisation in plants?

A

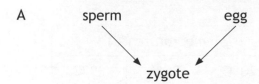

B

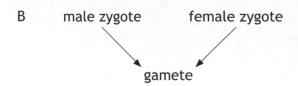

C

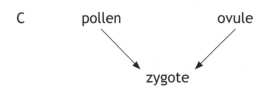

D

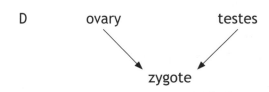

12. Variation in a characteristic can either be discrete or continuous. The range of heights and weights for a group of students were measured and recorded. Ear lobe types were also examined and categorised into groups.

Which line in the table below identifies the type of variation shown by each of these human characteristics?

	Height	Weight	Ear lobe types
A	continuous	continuous	discrete
B	discrete	continuous	continuous
C	discrete	discrete	continuous
D	continuous	discrete	discrete

13. The diagram below shows part of the human respiratory system.

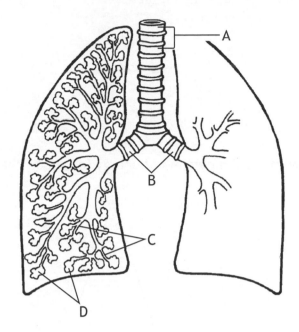

Which letter identifies the alveoli?

14. Which line in the table below identifies abiotic and biotic factors?

	Abiotic factor	Biotic factor
A	light intensity	pH
B	temperature	predation
C	grazing	light intensity
D	predation	grazing

[Turn over

15. A rabbit feeds on grass, is eaten by foxes and is a habitat for fleas.

   The statement above describes the rabbit's

   A   ecosystem

   B   community

   C   niche

   D   prey.

16. The diagram below shows the pyramid of energy for a food chain.

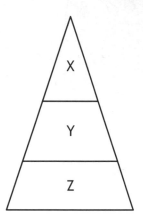

   There is a lot less energy at level X in the pyramid because

   A   there are fewer organisms at this level

   B   energy is stored at each level

   C   energy is lost at each level

   D   the organisms are bigger at this level.

17. In which of the following would competition **not** occur?

   A   Rabbits grazing in a field

   B   Owls and foxes hunting for mice

   C   Daisies and dandelions growing in a lawn

   D   Algae and fish in a loch

18. The following diagrams represent part of the nitrogen cycle. Which diagram shows the correct sequence of events in the nitrogen cycle?

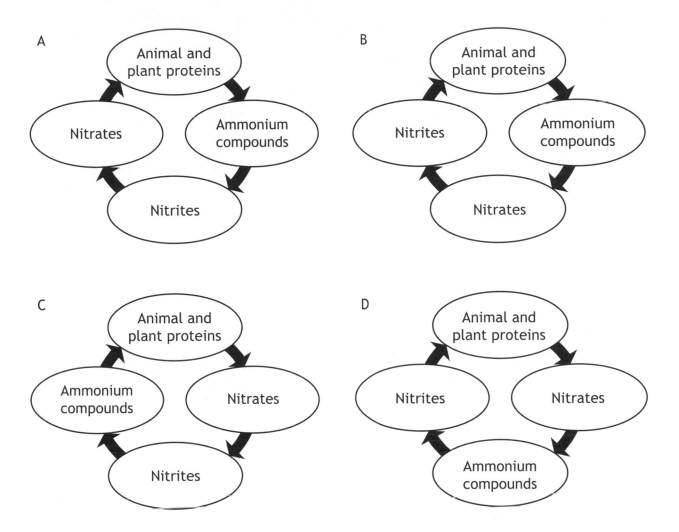

A

Animal and plant proteins

Nitrates

Ammonium compounds

Nitrites

B

Animal and plant proteins

Nitrites

Ammonium compounds

Nitrates

C

Animal and plant proteins

Ammonium compounds

Nitrates

Nitrites

D

Animal and plant proteins

Nitrites

Nitrates

Ammonium compounds

19. Students used a quadrat to estimate the number of buttercups in a field.

They threw the quadrat randomly three times in the area.

In order to improve the reliability of their results they could have

A   asked another group of students to check that they had counted correctly

B   thrown the quadrat ten times instead of three

C   only thrown the quadrat when conditions were at an optimum

D   used a smaller quadrat for each of their samples.

[**Turn over for Question 20 on *Page ten***

20. The table below compares the rate of extinction of mammal species over two different time periods.

Time period (years)	Rate of extinction per 100 years
1500 – 1900	4·5
1900 – 2000	90

The ratio of extinction rates between 1900 – 2000 compared to 1500 – 1900 is

A   1:20

B   1:2

C   2:1

D   20:1.

[END OF SECTION 1. NOW ATTEMPT THE QUESTIONS IN SECTION 2 OF YOUR QUESTION AND ANSWER BOOKLET]

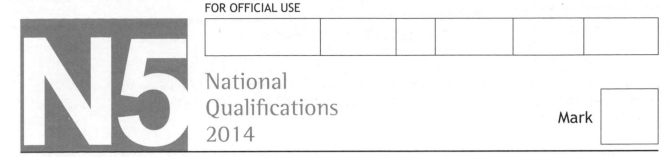

N5

National
Qualifications
2014

Mark

X707/75/01

Biology
Section 1—Answer Grid and Section 2

FRIDAY, 16 MAY

9:00 AM – 11:00 AM

Fill in these boxes and read what is printed below.

Full name of centre

Town

Forename(s)

Surname

Number of seat

Date of birth

Day	Month	Year
D D	M M	Y Y

Scottish candidate number

Total marks — 80

SECTION 1 — 20 marks

Attempt ALL questions in this section.

Instructions for the completion of Section 1 are given on Page two.

SECTION 2 — 60 marks

Attempt ALL questions in this section.

Write your answers clearly in the spaces provided in this booklet. Additional space for answers and rough work is provided at the end of this booklet. If you use this space you must clearly identify the question number you are attempting. Any rough work must be written in this booklet. You should score through your rough work when you have written your final copy.

Use **blue** or **black** ink.

Before leaving the examination room you must give this booklet to the Invigilator; if you do not, you may lose all the marks for this paper.

SECTION 1— 20 marks

The questions for Section 1 are contained in the question paper X707/75/02.
Read these and record your answers on the answer grid on Page three opposite.
Do NOT use gel pens.

1. The answer to each question is **either** A, B, C or D. Decide what your answer is, then fill in the appropriate bubble (see sample question below).

2. There is **only one correct** answer to each question.

3. Any rough working should be done on the additional space for answers and rough work at the end of this booklet.

Sample Question

The thigh bone is called the

> A   humerus
>
> B   femur
>
> C   tibia
>
> D   fibula.

The correct answer is **B**—femur. The answer **B** bubble has been clearly filled in (see below).

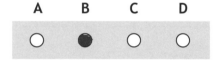

Changing an answer

If you decide to change your answer, cancel your first answer by putting a cross through it (see below) and fill in the answer you want. The answer below has been changed to **D**.

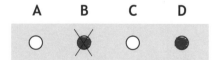

If you then decide to change back to an answer you have already scored out, put a tick (✓) to the **right** of the answer you want, as shown below:

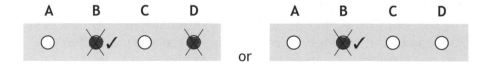

SECTION 1 — Answer Grid

	A	B	C	D
1	○	○	○	○
2	○	○	○	○
3	○	○	○	○
4	○	○	○	○
5	○	○	○	○
6	○	○	○	○
7	○	○	○	○
8	○	○	○	○
9	○	○	○	○
10	○	○	○	○
11	○	○	○	○
12	○	○	○	○
13	○	○	○	○
14	○	○	○	○
15	○	○	○	○
16	○	○	○	○
17	○	○	○	○
18	○	○	○	○
19	○	○	○	○
20	○	○	○	○

[BLANK PAGE]

DO NOT WRITE ON THIS PAGE

[Turn over for Question 1 on *Page six*

DO NOT WRITE ON THIS PAGE

SECTION 2 — 60 marks

Attempt ALL questions

1. A group of students carried out an investigation into the variety of cell types.

The types of cell they examined are shown in the box below.

Animal	Plant	Bacterial	Fungal

(a)   (i)   Identify the type(s) of cell which have a cell wall.   1

_____

(ii)   Identify the type(s) of cell which have a plasmid.   1

_____

(iii) Some organelles are found in all cells.

Choose one of the following organelles and tick (✓) the appropriate box.

Describe the function of the chosen organelle.   1

Ribosome ☐   Mitochondria ☐

Function _____

_____

MARKS

DO NOT
WRITE IN
THIS
MARGIN

1.   (continued)

(b) The students then measured a number of cells and calculated the average cell sizes. The results are shown in the table below.

Type of cell	Average size of cell (μm)
Animal	24
Plant	48
Bacterial	3
Fungal	7

On the graph paper below, complete the vertical axis and draw a bar chart to show the average size of the cells shown in the table.   2

(Additional graph paper, if required, can be found on *Page twenty-six*)

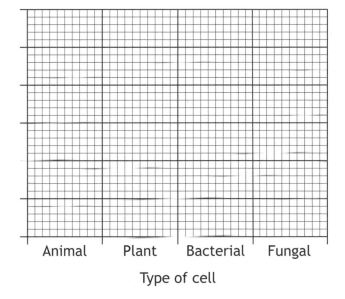

Animal   Plant   Bacterial   Fungal

Type of cell

Total marks   5

[Turn over

2. The apparatus shown below was used to investigate the movement of water into and out of a model cell. The model cell had a selectively permeable membrane.

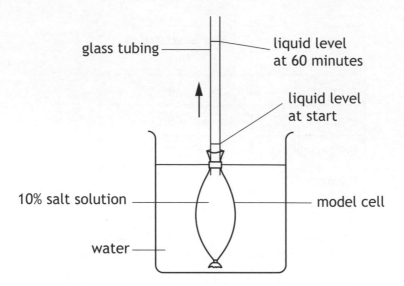

The liquid level in the glass tubing was measured every 10 minutes for 60 minutes.

The results are shown in the table below.

Time (minutes)	Liquid level (mm)
0	10
10	22
20	32
30	40
40	48
50	56
60	64

(a) Name the process which caused the liquid level to rise.   1

_____

MARKS | DO NOT WRITE IN THIS MARGIN

2. **(continued)**

(b) Explain how this process caused the liquid level to rise. 2

_____

_____

_____

(c) Calculate the average rate of movement of liquid in the glass tubing. 1

Space for calculation

_____ mm per minute

(d) When the investigation was repeated, the average rate of movement of liquid was slower.

Suggest **one** difference in the way that the investigation was set up that could have caused this change in results. 1

_____

_____

Total marks 5

[Turn over

MARKS | DO NOT WRITE IN THIS MARGIN

3. (a) Hydrogen peroxide can damage cells and lead to cell death. Catalase is an enzyme which breaks down hydrogen peroxide into oxygen and water.

Scientists in New Zealand investigated the link between the level of catalase in sheep livers and the fat in their meat. The hypothesis was that the higher the level of liver catalase, the greater the fat content of the meat.

In the investigation, they examined 9 sheep with a high percentage of fat and 15 sheep with a low percentage of fat. The sheep with the high percentage of fat had an average catalase level of 4800 K/g and those with the lower percentage of fat had an average catalase level of 3600 K/g.

The scientists concluded that their hypothesis was correct.

(i) Name the substrate of catalase.   **1**

_____

(ii) Identify an aspect in the planning of the investigation that would suggest that the hypothesis might not be proven correct.   **1**

_____

_____

(iii) A further investigation proved that the hypothesis was correct.

Describe how this investigation could help farmers to select only sheep with a low percentage of fat, to provide meat for consumers following a low fat diet.   **1**

_____

_____

(b) The optimum temperature for the activity of catalase is 37°C.

Predict what would happen to the activity of catalase if the temperature was lowered to 34°C.   **1**

_____

Total marks   4

MARKS

4. The following diagram shows a cross-section of some villi in the small intestine.

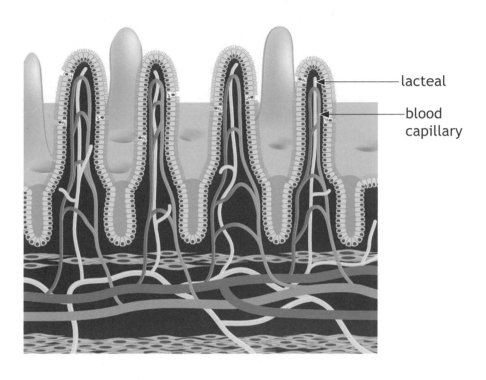

lacteal

blood capillary

Explain why the **structure and number** of villi make absorption an efficient process in the small intestine.

3

_____

_____

_____

_____

_____

_____

[Turn over

MARKS

5. Photosynthesis is a two stage process.

Stage 1 — Light reactions

Stage 2 — Carbon fixation

(a) The table below shows some statements about photosynthesis.

Complete the table to show which stage each statement refers to by placing a tick (✓) in the Stage 1 or Stage 2 box.

The first two statements have been completed for you.

2

Statement	Stage 1	Stage 2
Carbon dioxide required		✓
Light energy required	✓	
Water required		
Sugar produced		
ATP + Hydrogen required		
Oxygen produced		

(b) Explain why high temperatures (above 50°C) would prevent the photosynthesis reactions from taking place.

2

_____

_____

_____

5. (continued)

(c) The graph below shows how the rate of photosynthesis is affected by the concentration of carbon dioxide.

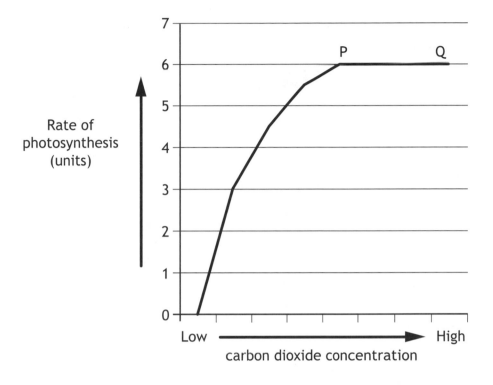

State two environmental factors which could limit the rate of photosynthesis between points P and Q.

1

1 _____

2 _____

Total marks   5

[Turn over

MARKS | DO NOT WRITE IN THIS MARGIN

6. The diagrams below show examples of some types of specialised cells from the human body.

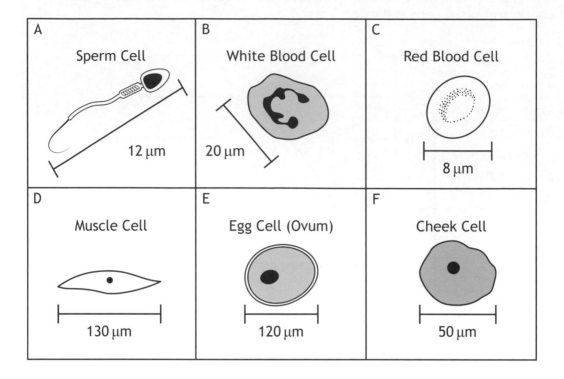

The cells are not drawn to the same scale.
(μm = micrometre)

(a) Put letters in the boxes below to arrange the cells in order of size.   1

increasing size

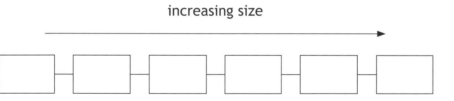

(b) Choose one of the following cell types by (circling) it.

sperm cell   egg cell   red blood cell

Describe the function of the chosen cell and explain how its specialisation allows it to carry out that function.   2

Function _____

Explanation _____

_____

Page fourteen

6. **(continued)**

(c) The diagram below shows some stages in the development of blood cells and nerve cells.

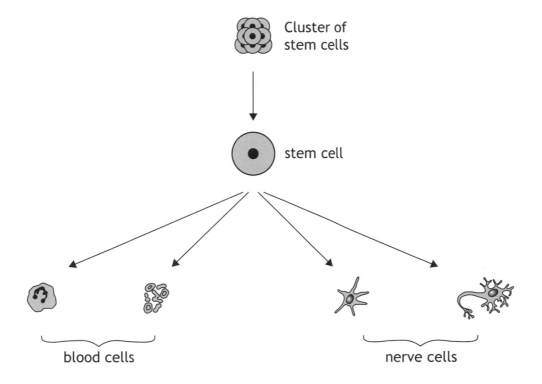

Cluster of stem cells

stem cell

blood cells

nerve cells

Describe the feature of stem cells which gives them the potential to develop into many different types of cells, such as blood and nerve cells. **1**

(d) Which of the following statements refer to processes involving stem cells?

Tick (✓) the correct box(es). **1**

Growth of new skin ☐

Transmission of nerve impulses ☐

Muscle contraction ☐

Repair of broken bones ☐

Production of insulin ☐

Total marks **5**

MARKS

7. Muscle tissue can be dark or light in colour.

Dark tissue cells use oxygen to release energy.

Light tissue cells do not use oxygen to release energy.

(a) Name the process by which energy is released in the dark tissue cells.

1

_____

(b) (i) Name the substance which muscle cells break down to produce pyruvate.

1

_____

(ii) When pyruvate is being formed, enough energy is released to form two molecules of a high energy compound.

Complete the word equation below to show how this compound is generated.

1

_____ + _____ ⟶ _____

(c) The table below shows the average percentage of dark and light tissue cells. These cells were found in the muscles of athletes training for different events at the 2014 Commonwealth games in Scotland.

Type of Athlete	Average percentage of dark tissue cells (%)	Average percentage of light tissue cells (%)
cyclist	60	40
swimmer	75	25
shot putter	40	60
marathon runner	82	18
sprinter	38	62

(i) Using information in the table, identify which type of athlete would be likely to produce the most lactic acid in their muscle cells. Justify your answer.

2

Type of athlete_____

Justification_____

_____

MARKS DO NOT WRITE IN THIS MARGIN

7. (continued)

(ii) A sample of muscle tissue from an athlete was examined and found to contain a total of 360 cells.

90 of these cells were light tissue cells.

Identify which type of athlete the sample was taken from. 1

Space for calculation

Type of athlete ______________________________

Total marks 6

[Turn over

MARKS

8. (a) The regulation of glucose in the blood is represented in the diagram below.

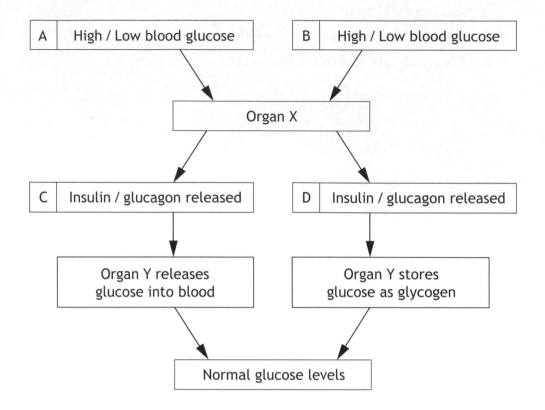

| A | High / Low blood glucose | B | High / Low blood glucose |

Organ X

| C | Insulin / glucagon released | D | Insulin / glucagon released |

Organ Y releases glucose into blood

Organ Y stores glucose as glycogen

Normal glucose levels

(i) The diagram above has two options in each of the four boxes A, B, C, D.

Circle the correct option in each box. 2

(ii) Identify organs X and Y. 2

Organ X _____

Organ Y _____

(b) Insulin and glucagon are hormones.

Describe two features of hormones. 2

1 _____

2 _____

Total marks 6

MARKS | DO NOT WRITE IN THIS MARGIN

9. Coat colour in Labrador dogs is an inherited characteristic. Black coat (**B**) colour is dominant to chocolate coat colour (**b**).

(a) A homozygous black Labrador was crossed with a Labrador with a chocolate coloured coat.

Complete the diagram below to show the genotypes of each of the parents and the F_1 phenotype.   2

Parents: black coat   X   chocolate coat

Genotypes: [   ]   [   ]

F_1 genotype: All Bb

F_1 phenotype: [   ]

(b) (i) Explain what is meant by polygenic inheritance.   1

_____

_____

(ii) State the type of variation shown by polygenic inheritance.   1

_____

Total marks   4

[Turn over

MARKS

10.　(a)　Lugworms live on the seashore in dark moist burrows under the sand.

The graph below shows the average number of lugworms at different distances from the seawater at low tide.

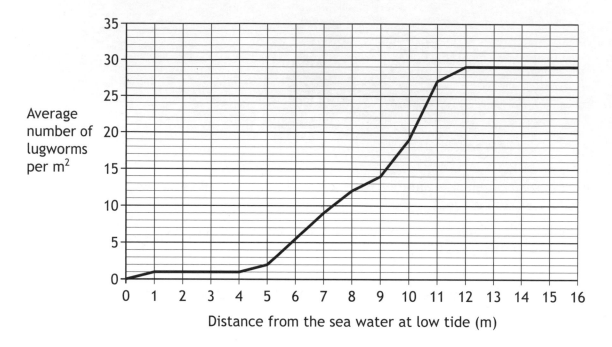

(i)　Describe the relationship between the distance from the seawater at low tide and the average number of lugworms per m^2.　**2**

_____

_____

_____

_____

(ii)　Calculate how many times greater the average number of lugworms at 11 metres is compared to 7 metres from the seawater at low tide.　**1**

Space for calculation

_____ times greater

MARKS | DO NOT WRITE IN THIS MARGIN

10.   **(continued)**

(b) Dover sole and rex sole are different species of flatfish and are predators of lugworms. Curlews, which are a species of wading bird, also feed on lugworms.

(i) Complete the table below by placing a tick (✓) in the correct box to show the type of competition that would occur between the different predators.   1

Predator	Type of Competition	
	Intraspecific	*Interspecific*
rex sole and curlew		
curlew and curlew		
rex sole and dover sole		

(ii) A curlew gains an average of 165 kilojoules (kJ) of energy daily, by feeding on lugworms.

Select, from the following list, the value of the energy which is used for growth each day by the curlew.

Tick (✓) the correct box.   1

165 kJ ☐

148·5 kJ ☐

16·5 kJ ☐

0 kJ ☐

Total marks   5

[Turn over

MARKS

11. During a woodland survey, a group of students measured some abiotic factors. Readings they took included the temperature of the soil and the air.

(a) Name one abiotic factor, other than temperature, which they could have measured in the woodland and describe the method of measuring this factor.   **2**

Abiotic factor _____

Method _____

_____

_____

(b) (i) During the survey, the students sampled the leaf litter in the woodland using pitfall traps.

However, when they checked the pitfall traps four days after setting them up, the students discovered that they were all empty.

Describe an error the students might have made which would explain why there were no invertebrates in the traps.   **1**

_____

_____

_____

MARKS

11. (b) (continued)

(ii) The error was corrected and the students set out the pitfall traps once again. The table below shows the types of invertebrates and numbers found.

Invertebrates	Number found
Woodlice	35
Beetles	20
Slugs	0
Spiders	30
Snails	15

Use the information in the table to complete the pie chart below.   2

(An additional pie chart, if required, can be found on *Page twenty-six.*)

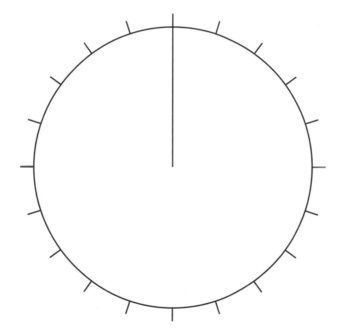

(c) The students saw a large number of butterflies in the woodland.

Give a reason why no butterflies were collected with the invertebrates.   1

_____

_____

Total marks   6

12. The following diagram shows the stages in the formation of a new species.

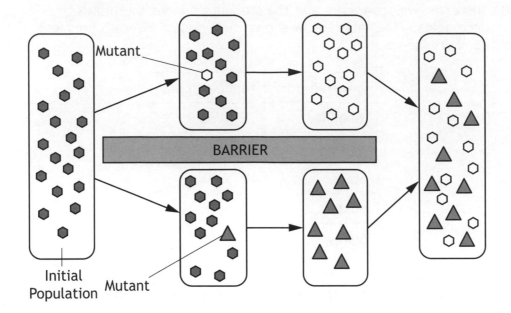

(a) Using the information in the diagram, describe how new species are formed.

4

_____

_____

_____

_____

_____

_____

_____

_____

_____

_____

_____

_____

MARKS | DO NOT WRITE IN THIS MARGIN

12.   (continued)

(b)   Choose either mutation or species and tick (✓) the appropriate box.
Give a definition of the chosen term.   **1**

Mutation ☐   Species ☐

Definition _____

_____

_____

(c)   In any population, variation exists. Explain why variation is important for the survival of a population.   **1**

_____

_____

Total marks   6

[END OF QUESTION PAPER]

ADDITIONAL SPACE FOR ANSWERS

ADDITIONAL GRAPH PAPER FOR QUESTION 1(b)

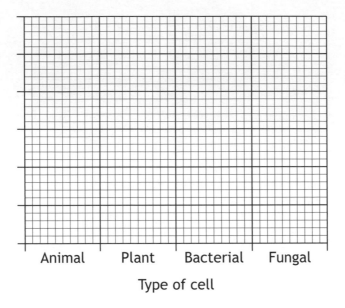

Type of cell

ADDITIONAL PIE CHART FOR QUESTION 11(b)

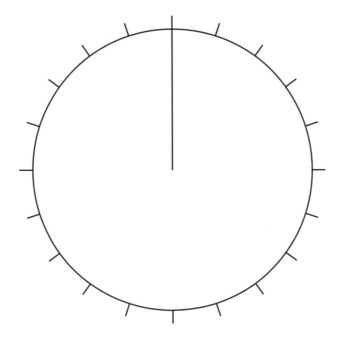

MARKS | DO NOT WRITE IN THIS MARGIN

ADDITIONAL SPACE FOR ANSWERS AND ROUGH WORK

ADDITIONAL SPACE FOR ANSWERS AND ROUGH WORK

Acknowledgement of Copyright

Questions 1, 4 and 9 images from www.shutterstock.com

SQA AND HODDER GIBSON NATIONAL 5 BIOLOGY 2014

NATIONAL 5 BIOLOGY SPECIMEN QUESTION PAPER

SECTION 1

Question	Response
1.	D
2.	A
3.	A
4.	A
5.	B
6.	C
7.	B
8.	A
9.	C
10.	D
11.	A
12.	D
13.	B
14.	D
15.	B
16.	B
17.	D
18.	A
19.	C
20.	C

SECTION 2

1. (a) Y

   (b) Large number of mitochondria present

   (c) Chloroplasts present = 1
   They contain chlorophyll/green pigment/are green = 1

2. (a) (i) From cell of alveolus wall to cell of capillary wall to plasma red blood cell
   (ii) Oxygen moves from a higher concentration to a lower concentration

   (b) Osmosis would not occur
   AND
   There is no concentration gradient/difference in concentration

3. (a) (i) mRNA
   (ii) 1. Bases = 1   2. C = 1

   (b) Amino acids

   (c) Different sequence/order of bases

4. (a) (i) Hydrogen
   (ii) Light energy is trapped by chlorophyll = 1
   Light energy/it is converted into ATP = 1
   ATP goes from stage 1 to stage 2 = 1

   (b) Light intensity
   Carbon dioxide concentration

5. (a) Carbon dioxide

   (b) pH 5
   Highest (average) number of bubbles (for most groups)

   (c) All flasks at same pH = 1

   Yeast — different types of yeast in each flask
   OR
   Temperature — different temperatures
   OR
   Glucose — different glucose concentrations used = 1

6. (a) (i) The hormone and receptor have complementary shapes/the hormone fits the receptor on the target cell only/target cell has specific receptors
   (ii) Hormone message — chemical, carried in blood, long-lasting, carried all over body
   Nerve message — electrical, short-lived effect, carried along specific nerves/path

   (b) (i) Type 1 — Insulin not produced
   (ii) Would stay higher than normal/would stay too high
   (iii) Pancreas

7. (a) Correct label copied exactly as in table and scale on y-axis (must use more than half of graph paper) = 1
   Correct plot joined with straight lines = 1

   (b) (i) Heart rate unchanged at low concentrations/0·25–0·75ppm = 1
   After that/from 0·75–2·00ppm heart rate decreases = 1

   (ii) Agree — it affected water fleas, so could affect humans/both humans and water fleas have hearts so both could be affected
   Disagree — water fleas and humans are very different/this is only one study/not enough data to predict
   OR
   any other justifiable answer

   (c) To allow comparison with the ones with the hormone/to show the effect that the hormone had

8. (a) Heterozygous

   (b) HH   Hh
   Hh   hh

   (c) HH or Hh

9. (a) Choose any two of arteries, veins and capillaries

   Comparison of:
   Thickness of walls
   Muscularity of walls
   Presence and absence of valves
   Size of channel for blood flow
   Any 3 correct

   (b) Carries oxygen

   (c) (i) Although country B has fewer dying from heart disease, more of them die from childhood diseases/infectious diseases
   OR
   Only one part of the data has been considered and the whole of the information needs to be taken into account

(ii) Stop smoking/avoid smoke-filled environments/ take more exercise

10. (a) (i) Mouths are all different shapes/sizes/structures
        (ii) A, B , D (**Any two**) — they all rely on fish for food

    (b) Niche

    (c) (3)-1-5-4-2

    (d) All the organisms living in a particular area and the non-living components (with which they interact)

11. (a) L

    (b) Biomass — shows the total mass of living organisms/ population at each stage/level in a food chain

    (c) (i) 9700
        (ii) Heat
            OR
            Movement
            OR
            Undigested material

12. (a) (i) As the mass of nitrogen fertilizer increases, the average mass of root nodules decreases rapidly until 0·4g/1g = **1**
            Then decreases more slowly = **1**
        (ii) Any mass < 0·25g (including 0)

    (b) Nitrifying bacteria — converts ammonium compounds into nitrites and/then nitrates
        OR
        Denitrifying bacteria — converts nitrates in soil into nitrogen gas in the air
        OR
        Root nodule bacteria — fix nitrogen in the air into nitrates

13. (a) (i) Initial populations all had different starting sizes
        (ii) 4·3
        (iii) Starling and yellow wagtail

    (b) Biological control — explanation should relate to the third, fifth or sixth items on the list
        Explanation = **1**
        Further explanation = **1**
        OR
        GM crops — explanation should be related to the third, fourth, fifth or sixth items on the list
        Explanation = **1**
        Further explanation = **1**
        (Marks are for explanation only. Two explanations from three are acceptable for 2 marks. Answers must state **how** some of the disadvantages are overcome.)

NATIONAL 5 BIOLOGY
MODEL PAPER 1

SECTION 1

Question	Response
1.	C
2.	B
3.	D
4.	D
5.	B
6.	A
7.	C
8.	A
9.	B
10.	C
11.	D
12.	D
13.	A
14.	C
15.	A
16.	A
17.	D
18.	B
19.	C
20.	B

SECTION 2

1. (a) (i) 1 lipid
            2 protein
            1 each
        (ii) F
            C
            E
            All = 2, 2/1 =1

    (b) 400 micrometres

2. (a) (i) Spindle fibre
        (ii) Chromatids arrive at poles/ends of spindle
            OR
            Two new nuclei form
            OR
            Cytoplasm splits

    (b) 17%

    (c) Two matching sets
        Diploid
        Both = 1

3. (a) (i) Scales and labels = **1**
            Plotting and line connection = **1**

    (b) 35°C

    (c) Ensures that reaction takes place at the intended temperature

    (d) Enzymes are specific to one substrate only

    (e) Repeat the investigation at 35°C/constant temperature
        AND
        Repeat at different pH levels

4. (a) (i) Light reactions/light dependent stage
    (ii) Oxygen
    (iii) X – hydrogen/hydrogen attached to acceptor molecule
    AND
    Y – ATP
    (iv) Combines with hydrogen to produce sugar

  (b) A – light intensity
    B – CO_2 concentration
    C – temperature
    All = 2, 2/1 = 1

5. (a) (i) Primitive blood cell
    (ii) Basic bone cell

  (b) (i) Meristems
    (ii) Xylem, phloem, mesophyll, epidermis
    Any 2 = 1

6. (a) Cerebrum – memory and reasoning
    Cerebellum – control of balance
    Medulla – control of heart rate
    All 3 = 2, 2/1 =1

  (b) Sensory
    Relay
    Motor
    All 3 = 2, 2/1 = 1

  (c) Protection

7. (a) A – sperm cell
    B – egg cell/ovum
    Both = 1

  (b) A – testes
    B – ovary
    Both = 1

  (c) Containing one set of chromosomes

  (d) Zygote

8. (a) (i) Water moves into root cells by osmosis = **1**
    From higher concentration in soil to lower concentration in cells = **1**
    Root hair cells increase the surface area of root for osmosis = **1**

    (ii) Lignin

  (b) Photosynthesis
    OR
    Support
    OR
    Cooling

9. (a) 2
    6
    3
    7
    All 4 = 2, 3/2 = 1

  (b) Thicker
    Narrower
    High
    Away from
    All = 2, 3/2 = 1

  (c) Haemoglobin

10. (a) (i) B, F **both**
    (ii) A, D **both**
    (iii) C **OR** H

  (b) Variety and relative abundance of living organisms

  (c) Deforestation, habitat destruction, pollution, overfishing/hunting/grazing/intensive farming/others possible
    Any one

11. (a) 7

  (b) 2 : 3

  (c) Lip flush with soil — allows animals to enter
    Empty regularly — avoid predation
    Provide drainage holes — avoid drowning
    Any:
    Precautions = 1
    Matching reasons = 1

  (d) Sectors = **1**
    Labelling = **1**

12. (a) Mutation is random changes to organism's genotype/genes/genetic material = **1**
    (Beneficial) mutation gives selective advantage which allows survival = **1**
    Mutation/favourable characteristic passed to offspring = **1**

  (b) Reduces competition

  (c) Speciation

NATIONAL 5 BIOLOGY MODEL PAPER 2

SECTION 1

Question	Response
1.	B
2.	A
3.	C
4.	A
5.	D
6.	C
7.	A
8.	B
9.	B
10.	D
11.	A
12.	D
13.	D
14.	B
15.	C
16.	A
17.	B
18.	C
19.	D
20.	C

SECTION 2

1. (a)　(i) X – cell wall
　　　　　Y – plasmid
　　　　　1 mark each
　　　　(ii) Site of protein synthesis

　(b) Difference – walls made of different substances/ fungi have organelles/nucleus but bacteria do not = **1**
　　　Similarity – both have cell walls/cytoplasm/genetic material/membranes/ribosomes = **1**

2. (a) Labels = **1**
　　　Plotting and line connection = **1**

　(b) Temperature/volume of salt solution /shape/surface area of potato
　　　Any 2 = 1

　(c) 5g/100cm³ = **1**
　　　Little or no change expected in the final weight of tissue placed in this concentration = **1**

3. (a) Nucleus

　(b) Double stranded/double helix = **1**
　　　Adenine/thymine/guanine/cytosine bases named = **1**
　　　Bases on each strand form complementary pairs = **1**

　(c) Enzymes/hormones/antibodies/structural units/ receptors
　　　any 2

4. (a) Step 2 human gene extracted, plasmid removed = **1**
　　　Step 3 plasmid opened and human gene sealed in = **1**
　　　Step 4 transformed plasmid inserted into bacterial cell = **1**

　(b) Insulin/growth hormone

　(c) Fertilisation
　　　OR
　　　By a virus/vector
　　　OR
　　　By conjugation

5. (a) X – carbon dioxide = **1**
　　　Y – water = **1**

　(b) Aerobic – 38
　　　Fermentation – 2
　　　Both = 1

　(c) Cytoplasm

6. (a) Pancreas
　　　(i) X – insulin = **1**
　　　　　Y – glucagon = **1**
　　　(ii) Glycogen = **1**
　　　　　Liver = **1**
　　　(iii) Sedentary lifestyles

　(b) Obesity
　　　Sugar/fat-rich diet
　　　Increased life expectancy/population
　　　Any 1

7. (a) Accurate sectors = **1**
　　　Labelling OR key = **1**

　(b) 30%

　(c) Continuous – variation which shows a range/can be measured/is not clear-cut = **1**
　　　Polygenic – controlled by alleles of more than one gene = **1**

8. (a)　(i) 12 hours
　　　　(ii) Between 3pm and 6pm
　　　　(iii) 9am

　(b) 1 Water absorbed by osmosis into root (hair) cells
　　　2 Water passed through stems in xylem (vessels)
　　　3 Water evaporated from (leaf) mesophyll cells
　　　4 Transpiration through stomata (in epidermis)
　　　Any 3, 1 mark each

9. (a)　(i) Pulmonary artery
　　　　(ii) Aorta
　　　　(iii) Coronary artery

　(b) Regular exercise/low fat diet/healthy diet not smoking/others
　　　Any 2, 1 each

10. (a) Uptake 5
　　　Decomposition 2
　　　De-nitrification 6
　　　Death 1
　　　All 4 = 2, 3/2 = 1

　(b) Nitrate

　(c) Nitrifying

　(d) Making amino acids/polypetides/protein/nucleic acid

11. (a) 290kg

(b) 140%

(c) To show the effects of the genetic modification on cotton yield
OR
To compare with the genetically modified cotton plants

(d) Adding genetic resistance to boll weevil/pests increases the yield of cotton plants
OR
GM crops give higher yields
OR
Boll weevil/pest can reduce yield of cotton crops
Any = 1

(e) High yields without the need for pesticide control
OR
Less cost for /environmental impact of pesticide

12. (a) (i) Decreases
(ii) Increases

(b) High levels of SO_2
OR
Acidic/low pH rain

(c) $20\mu g/m^3/km$

NATIONAL 5 BIOLOGY MODEL PAPER 3

SECTION 1

Question	Response
1.	A
2.	C
3.	B
4.	C
5.	B
6.	D
7.	D
8.	A
9.	C
10.	B
11.	B
12.	A
13.	B
14.	D
15.	D
16.	C
17.	C
18.	A
19.	D
20.	A

SECTION 2

1. (a) (i) Osmosis

(ii) Water moved from higher concentration inside the cells to lower concentration in the sugar solution

(b) Substance – oxygen/glucose
OR
CO_2 = 1
Reason – required for respiration
OR
Waste removal = **1 must match**

(c) Active transport requires additional energy passive does not
OR
Active is against concentration gradient passive is down the gradient
OR
Active involves carrier protein passive does not

2. (a) Steady for two/three hours
Increases up to 14 hours
Remain steady
All 3 = 2, 2 = 1

(b) 10–12 hours

(c) 200%

(d) 1:9

(e) To prevent contamination by unwanted species/ maintain sterility

3. (a) 4.2

   (b) Keep volume/concentration of enzyme
   OR
   Depth of agar/gel/well the same for each enzyme
   OR
   Diameter of well
   OR
   pH of agar/gel

   (c) Specificity of enzyme activity
   OR
   Enzymes only act on one substrate

   (d) Speed up chemical reactions in living cells

   (e) High temperature/extremes of pH

4. (a) Chlorophyll

   (b) ATP

   (c) Hydrogen

   (d) Starch
   OR
   Cellulose

   (e) As temperature decreases, it slows down enzymes needed for photosynthesis

5. (a) (i) 10%
   (ii) Increasing the concentration of ATP increases the contraction it causes in muscle cells
   (iii) Start lengths were different

   (b) Prevent cross contamination of ATP solutions

   (c) Aerobic – X, Y
   Fermentation – W, X, Z
   All = 2, 1 error = 1

6. (a) A
   B
   D
   F
   All 4 = 1

   (b) Synapse

   (c) Chemical diffuses across the gap

7. (a) (i) A – ovary
   B – anther
   1 mark for each
   (ii) Fertilisation
   (iii) Zygote

   (b) Ovule – haploid
   Pollen – haploid
   Cell Y – diploid
   All 3 = 2, 2 = 1

8. (a) (i) Male 1 – Tt
   Female 2 – tt
   Female 4 – Tt
   All 3 = 2, 2/1 - 1
   (ii) Tick male parent only

   (b) Alleles

   (c) Does not show a range
   OR
   Falls into clear cut categories

9. (a) **Artery**
   Structure – thick-walled/elastic walls/
   narrow internal diameter (**Any 2**)
   Function – blood under high pressure/
   transport blood from heart/
   most carry oxygenated blood (**Any 1**)
   OR
   Vein
   Structure – thinner walled/wider internal diameter/
   have valves (**Any 2**)
   Function – blood under low pressure/
   transport blood to the heart/
   mostly carry deoxygenated blood (**Any 1**)
   OR
   Capillary
   Structure – one cell thick walls/thin walls/
   narrow diameter/large surface area (**Any 2**)
   Function – blood under reducing pressure/
   exchange materials with tissues,/allow named
   substance to pass through (**Any 1**)

   (b) (i) A group of similar cells which carry out the same function
   (ii) Stem cells

10. (a) Genetically modified/GM/genetically engineered

    (b) Chemicals not used **OR** control achieved through use of the pest's natural parasites/predators/diseases

    (c) 1 Leaches/washes into fresh water
    2 Promotes growth of algae/increases algal bloom
    3 Algae rotted by (aerobic) decomposer bacteria
    4 Water deoxygenated
    Any 3, 1 mark each

11. (a) (i) X axis scales and label = 1
    plotting and connection = 1
    (ii) 12 times

    (b) Sample indicator organisms above and below sewage discharge = 1
    Compare samples to show effects of sewage on river = 1

12. (a) Lack of competition for food/space
    OR
    Low incidence of disease
    OR
    Few predators

    (b) Limited space/food, increased disease/competition for resources

    (c) (i) Random change to genetic material
    (ii) Survival of the fittest/natural selection

NATIONAL 5 BIOLOGY 2014

SECTION 1

Question	Response
1.	B
2.	D
3.	B
4.	C
5.	D
6.	C
7.	A
8.	C
9.	A
10.	A
11.	C
12.	A
13.	D
14.	B
15.	C
16.	C
17.	D
18.	A
19.	B
20.	D

SECTION 2

1. (a) (i) Plant, bacterial and fungal
   (ii) Bacteria(l)
   (iii) Ribosome – (site of/involved in) protein synthesis
   Mitochondria – (site of/involved in) energy or ATP production/aerobic respiration

   (b) Y axis scale and label, including units
   Bars correctly plotted

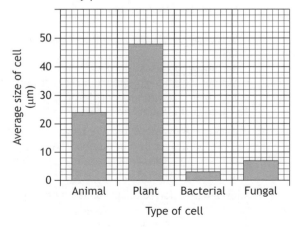

2. (a) Osmosis
   (b) Water moves into the (model) cell/bag/salt solution   1 mark
   From a high water concentration to a low water concentration/down a concentration gradient 1 mark
   OR
   Alternative answer for 2 marks:
   Water moves from a high water concentration outside to a low water concentration inside the (model) cell/bag/ salt solution
   (c) 0.9
   (d) Description of concentration change – must be a smaller concentration gradient than shown/lower temperature/wider capillary tube/seal not tight/ less water in the beaker/bag not fully submerged

3. (a) (i) Hydrogen peroxide
   (ii) Numbers (in each group) different
   OR
   Overall numbers used too small
   (iii) If they have a low level of catalase/only use sheep with low levels of catalase/don't use sheep with high levels of catalase

   (b) (Activity is) decreased/slows down reaction

4. A villus has:
   A thin wall
   A large surface area
   A good blood supply/many capillaries

   • There are a large number of villi
   • So this also increases surface area/creates a large surface area
   OR
   makes absorption/diffusion fast(er)/more

5. (a)

Statement	Stage 1	Stage 2
Carbon dioxide required		✓
Light energy required	✓	
Water required	✓	
Sugar produced		✓
ATP + Hydrogen required		✓
Oxygen produced	✓	

   1 mark for each correct column

   (b) Photosynthesis is controlled by enzymes/enzymes are needed
   (At high temperatures) enzymes are denatured/do not work.

   (c) Light Intensity ⎫ either order
   Temperature ⎭

6. (a) C A B F E D
   (b) (Sperm)
   F Reproduction/fertilisation (or correct description)
   E Tail to swim/mitochondria for energy/haploid to allow fusion at fertilisation producing a diploid zygote
   OR
   (Egg cell)
   F Reproduction/fertilisation (or correct description)
   E Yolky cytoplasm/large cell to provide food or haploid to allow fusion at fertilisation producing a diploid zygote

OR

~~Red blood cell~~

F Carry oxygen
E Contain haemoglobin/carries oxygen as oxyhaemoglobin
OR
large surface area/ biconcave/no nucleus to transport more oxygen
OR
small/flexible to go through capillaries

(c) Unspecialised/undifferentiated/not specialised

(d)

Growth of new skin	✓
Transmission of nerve impulses	☐
Muscle contraction	☐
Repair of broken bones	✓
Production of insulin	☐

7. (a) Aerobic respiration

(b) (i) Glucose
(ii) ADP + Phosphate/Pi/ PO_4 ⟶ ATP

(c) (i) Sprinter

Highest lactic acid produced when oxygen is not used to release energy
OR
Highest percentage light tissue
OR
Highest fermentation
OR
Highest percentage of cells that do not use oxygen.
(ii) Swimmer

8. (a) (i) A = low   B = high
        C = glucagon   D = insulin

(ii) Organ X = pancreas
     Organ Y = liver

(b) *Any two features from:*
• Made of protein
• Chemical messengers
• Specific for some (target) tissues
• Shaped to fit receptors
• Released or produced by endocrine glands/ system
• Carried in blood
• Can have a long term effect

9. (a) Genotypes: BB and bb
        F_1 phenotype: black (coat)

(b) (i) More than one/several genes control one/a characteristic
(ii) Continuous

10. (a) (i) • (The general trend is) as the distance increases, numbers/population/lugworms increases up to 12 metres
        • After that, numbers/population remains steady/stays the same
(ii) 3

(b) (i)

Predator	Type of Competition	
	Intraspecific	Interspecific
rex sole and curlew		✓
curlew and curlew	✓	
rex sole and dover sole		✓

(ii) 16.5 kJ

11. (a) Named abiotic factor, eg moisture, pH, light intensity
Description of method, including instrument name, eg "Stick the probe of the pH meter in the soil"

(b) (i) Left traps too long/
Traps too high above soil/
Traps not camouflaged/
Traps too shallow

(ii)

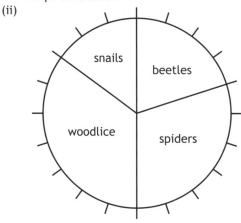

1 mark for appropriate sized sections
1 mark for labels

(c) Able to fly/flew away

12. (a) • Initial population is separated/split (or idea of this)
• (Different) mutations occur in each subpopulation/group (need indication that it is more than the original one population)
• Some mutations are advantageous
• Natural selection occurs
OR
Selection pressures are different in each group
OR
Advantageous mutations are selected for
• Subpopulations/groups are no longer able to interbreed to produce fertile offspring
Any two from last three bullet points

(b) Mutation – a (random) change to genetic material/ chromosome structure or number/bases in DNA
Species – organisms which can interbreed/reproduce to produce fertile offspring

(c) Allows population to adapt to changing environmental conditions
OR
suitable example of coping with change
OR
makes it possible for population to evolve in response to changing conditions

Acknowledgements

Permission has been sought from all relevant copyright holders and Hodder Gibson is grateful for the use of the following:
Image © Lebendkulturen.de/Shutterstock.com (SQP Section 2 page 15);
Image © Margarita Borodina/Shutterstock.com (SQP Section 2 page 17);
Image © Darren Baker/Shutterstock.com (2014 Section 2 page 6);
Image © Blamb/Shutterstock.com (2014 Section 2 page 11);
Image © Ysbrand Cosijn/Shutterstock.com (2014 Section 2 page 19).

Hodder Gibson would like to thank SQA for use of any past exam questions that may have been used in model papers, whether amended or in original form.